FIELD AND THEORY

FIELD and THEORY

Lectures in Geocryology

TERRAIN et THÉORIE

Essais de géocryologie

ПОЛЕВЫЕ РАБОТЫ И ТЕОРИЯ

ДОКЛАДЫ ПО ГЕОКРИОЛОГИИ

Edited by
Michael Church
Olav Slaymaker

University of British Columbia Press
Vancouver
1985

Field and Theory
Lectures in Geocryology

© The University of British Columbia 1985

Canadian Cataloguing in Publication Data

Main entry under title:

Field and theory

Lectures presented at a lecture series held at
the University of British Columbia during
1980-81.
Abstracts in French and Russian.
Bibliography
ISBN 0-7748-0204-9

1. Frozen ground — Addresses, essays, lectures.
I. Church, Michael Anthony, 1942-
II. Slaymaker, Olav, 1939-
III. Title: Terrain et théorie.
GB642-F53 1985   551.3'84   C84-091417-2

Printed in Canada on acid-free paper

ISBN 0-7748-0204-9

CONTENTS

ОГЛАВЛЕНИЕ

FIGURES AND ILLUSTRATIONS

TABLES

PREFACE

At the end of June, 1981, Dr. J. Ross Mackay retired from his position as Professor of Geography in the University of British Columbia. The date is of little consequence in terms of his continuing career of inquiry, innovative thinking and scholarly achievement, but it marks the end of a chapter of academic service that his colleagues were eager to celebrate. Accordingly, a series of lectures by internationally known scientists was held in the university during 1980-81 and these, plus contributions from some of Mackay's former graduate students, form the present monograph. Each of the authors represents a group of scholars with whom Mackay has contact.

The papers presented here illustrate well the critical nature of Ross Mackay's contribution to geocryology while they delineate a central dilemma in the field. The dilemma concerns the intricate challenge of field work in the harsh periglacial environment and the resulting difficulty of testing theory in the field with rigour.

In this monograph papers on soil freezing, ice formation and thaw at the microscale present a relatively sophisticated, quantitative treatment firmly based in theory. Computational methods and extension of results to engineering site evaluations are given in other contributions. However, the reliance upon one dimensional arguments in all of this work leads to difficulties in field applications. In comparison, regional accounts of geocryological and nival phenomena remain entirely empirical and mainly qualitative. This critical mismatch of understanding between microscale and regional scale is emphasized over and over by A. L. Washburn in his review of the status of periglacial studies. Alone amongst contemporary students, Ross Mackay has demonstrated consistently how to occupy the middle ground. His papers represent a sustained demonstration of the application of simple physical concepts to explain variations in the landscape. That represents his most valuable contribution to science.

M.C., O.S.

PREFACE

A la fin de juin 1981, le Professeur J. Ross Mackay prenait sa retraite du Département de géographie de l'Université de Colombie Britannique. Cette date ne marque pourtant pas une étape importante en termes de sa carrière de chercheur scientifique, d'innovateur et d'académique, mais elle souligne la fin d'une période de contribution académique que ses collègues ne pouvaient pas laisser passer inaperçue. En conséquence, une série de conférences par des scientifiques de réputation internationale fut organisée à l'Université en 1980-81. Ces conférences, jointes à des communications de quelques uns des anciens étudiants du Professeur Mackay, constituent la présente monographie. Chaque auteur représente un groupe de savants avec qui le Professeur maintient des contacts.

Les articles présentés ici illustrent bien l'importance de la contribution du Professeur Mackay à la géocryologie, en plus de définir un problème particulier à ce domaine d'étude. Ce problème résulte d'une part du défi que posent les travaux sur le terrain dans le rude environnement à climat périglaciaire et d'autre part des difficultés inhérentes à toute vérification rigoureuse d'une théorie sur le terrain.

Dans cette monographie on présente des études de processus à échelle réduite sur l'engel du sol, et sur la formation et la fonte de la glace qui témoignent d'une approche quantitative relativement avancée et solidement ancrée dans la théorie. Des communications sont aussi présentées sur l'utilisation des méthodes quantitatives et l'application des résultats à l'évaluation des sites pour des travaux de génie. Cependant, en ne s'appuyant que sur des arguments unidimensionnels dans ce type de travail, on aboutit à des difficultés d'application sur le terrain. Par comparaison, les travaux régionaux sur les phénomènes géocryologiques et nivaux demeurent entièrement empiriques et presque toujours qualitatifs. A. L. Washburn dans son compte rendu sur l'état des études périglaciaires souligne à plusieurs reprises ce sérieux décalage de nos connaissances des phénomènes régionaux par rapport à nos connaissances des processus étudiés à échelle réduite. Seul parmi les scientifiques contemporains Ross Mackay a démontré avec consistance comment combler cette lacune. Ses travaux démontrent de façon continue que l'application de concepts physiques simples suffit à expliquer des variations de paysage. C'est là sa contribution scientifique la plus importante.

Traduit par D. A. St. Onge
et J. Veillette

ПРЕДИСЛОВИЕ

В конце июня 1981 г. профессор доктор Дж. Росс МАКАЙ, член Королевского (научного) Общества Канады, согласно административным правилам, ушел в отставку с положения Профессора Географии в университете Британской Колумбии. В смысле его продолжающейся деятельности в науке, его новаторского мышления и его достижений в науке, эта дата имеет малое значение. Тем не менее, она отмечает заключительный момент его педагогической деятельности и его коллеги решили это отметить. В университете, в течение 1981 г. состоялась серия докладов. Читали всемирно известные специалисты. Эти доклады и статьи представленные бывшими аспирантами проф. Макая выходят в настоящей книге. Каждый из авторов представляет группу ученых с которыми проф. Макай поддерживает научный контакт.

Доклады напечатанные в книге хорошо иллюстрируют большое значение вклада, который проф. Макай сделал в геокриологию, а также определяют основные проблемы этой науки; те сложные и требовательные задачи полевых работ, которые необходимо проводить в суровых перигляциальных условиях, а также, исходящие из этого трудности, встречаемые при попытках строго проверять теоретические формулировки в полевых условиях.

В настоящей книге группа докладов относительно замерзания пород, образования льда и его оттаивания в микромасштабах представляет собой довольно усовершенствованный количественный подход, стойко обоснованный в теории. Методы расчетов и применение результатов при оценке стройплощадок для инженерных работ подаются в некоторых других докладах. Однако, полагаясь на одноразмерное, в смысле масштаба, толкование во всех докладах этой группы, авторы стоят перед затруднениями при применении полученных результатов в практике. С другой стороны, региональные характеристики геокриологических и нивальных явлений остаются совершенно эмпирическими и в основном качественными. Это критическое несовпадение подходов в микро и региональной шкалах многократно подчеркивается др. А. Л. Вашбурном в его обзоре положения в перигляциальных науках. Среди современных ученых один только проф. Макай последовательно показывает как придерживаться посредней позиции. Его работы убедительно демонстрируют как можно применять простые понятия физики для выяснения перемен в ландшафте. И это является его самым важым вкладом в науку.

Translated into Russian by J. Solecki

ACKNOWLEDGEMENTS

This book, and the lectures upon which it has been based, have been made possible by grants from the H. R. Macmillan Family Foundation and from the Leon and Thea Koerner Foundation of Vancouver, British Columbia, from the Arctic and Alpine Research Committee and the Faculty of Arts in the University of British Columbia.

Dr. Robert H. T. Smith and Dr. John D. Chapman, successive Heads of the University of British Columbia Department of Geography, helped to plan and organize the lectures and made administrative services available. The cooperation of the editors of the University of British Columbia Press has been a great help.

Special acknowledgements must be made to Professor Jan Solecki, of the University of British Columbia Department of Slavonic Studies, for the Russian abstracts, to Dr. Denis St. Onge of the Geological Survey of Canada, for reviewing French translations, and to the reviewer who provided additional information about the history of the Schefferville 'bedstead'.

Most of all, we are grateful to the authors for their willing cooperation in preparing their manuscripts and in responding to editorial suggestions.

M.C., O.S.

NOTATION

A	area of avalanche starting zone	S	maximum runout distance (of avalanche), displacement
a	pile radius, cross sectional area (of avalanche)	S	entropy per mole
A,a,B	ice flow law coefficients (temperature dependent)	s	interfacial surface tension
A,B,C	reference points	T	temperature
C	volumetric heat capacity	t	time
C_a	apparent heat capacity	V	molar volume
C_m	volumetric heat capacity of soil-icewater mixture	V_p	saturated pore space
d	hydraulic depth of avalanche	V_w	specific volume of water
E_m	$h' + v^2/4g$ = avalanche specific energy	v	velocity
		W	Schaerer velocity parameter for avalanches
e	surface energy	w	top width of avalanche
F	segregation index	X	horizontal (distance) displacement of avalanche
g	acceleration of gravity	Y	vertical (distance) displacement of avalanche
H	heave (length)	x,z	Cartesian coordinates
h	film thickness	Z	depth
h'	vertical height of avalanche flow	Z_f	depth of frost penetration
J	drag coefficient	Z_{th}	depth of thaw plane
K	thermal conductivity	α ⎫	avalanche model parameters (constants)
k	hydraulic conductivity	β ⎭	
L	latent heat of fusion of water	$\gamma, \gamma_a, \gamma_s$	specific weight (air, snow)
L	pile length, step length	Δ	finite difference
M	avalanche mass	ϵ	shear strain ($\epsilon_e, \epsilon_s, \epsilon_0$)
n	step index	ϵ_s	secondary shear strain
P	pressure, weight "wetted' perimeter (of avalanche track)	ζ	temperature dependent parameter
Q	heat flux	θ	slope of avalanche path
q	water flux	Θ	slope angle, coil moisture content
R	thermal resistance, hydraulic radius of flow	μ	chemical potential, coefficient of sliding friction
R	gas constant	ξ	coefficient of turbulent friction
r	radius, radius of curvature	ρ	density
r_p	minimum pore size	σ	deviatoric stress (σ_e)
		σ_n	buoyant stress

τ	shear stress (τ_0, τ_a)
ϕ	average gradient of avalanche path
$\chi(\zeta)$	stress partition function
ψ	suction potential
ψ_t	total soil water potential

repeated subscripts

a, b, c	parameters at reference points A, B, C
a	air
d	disjoining pressure
f	frozen
f*	at freezing point
e	effective
i	ice phase
l	fluid in film
o	reference state
s	spreading pressure
t	total
u	unfrozen
v	vapour phase
w	water

1
ON THE SCIENTIFIC METHOD
OF J. ROSS MACKAY

W.H. Mathews

Department of Geological Sciences, The University of British Columbia
Vancouver, British Columbia, V6T 1W5 Canada

ABSTRACT

'Scientific method' properly includes all the endeavours from the conception of an idea until final presentation and judgement upon it. The formal conceptual aspects of this work are well known. In the field sciences, however, the qualities and characteristics of the scientist may be particularly significant in determining the success of work. In this paper, I consider some of the qualities that underlie J. R. Mackay's achievements, particularly his ability to define problems from field observation and to design and execute field investigations.

RÉSUMÉ:
Sur la méthode scientifique de J. Ross Mackay

'La méthode scientifique' comprend toutes les étapes depuis la conception d'une idée jusqu'à la présentation finale et son évaluation. Les aspects formels et abstraits de cette méthode sont bien connus. Cependant, dans les sciences du terrain, les qualités et les caractéristiques du scientifique peuvent être particulièrement importantes dans la détermination du succès du travail. Dans cet article, nous examinons quelques unes des qualités qui expliquent les succès de J. R. Mackay, en particulier, sa capacité pour circonscrire les problèmes à partir d'observations de terrain et, à définir et à mener à terme des enquêtes sur le terrain.

НАУЧНАЯ МЕТОДИКА ПРИМЕНЯЕМАЯ ДЖ. РОСС МАКАЕМ

В. Х. МАТЬЮС

РЕЗЮМЕ

‹‹Научная методика›› в сущности должна воплощать все усилия, от зарождения идеи до ее окончательного представления и оценки. Формальные стороны понятия такой работы всем хорошо известны. Однако, в технических науках, качества и особенности самого ученого могут оказаться особенно важными, предопределяя успех в его работе. В настоящем докладе рассматриваются те некоторые особенности, которые предопределяют успехи в работе Дж. Росс Макая. Это особенно относится к его умению определять проблемы при полевых наблюдениях, а также устанавливать и проводить полевые исследования.

The term "scientific method" properly includes all the endeavours from the conception of an idea to its full-grown presentation in publication. Observation and measurement, design and execution of experiments in the field and in the laboratory, later analyses of the results together with the addition of data and concepts from other branches of science, manipulation of data with statistics and computer modelling, followed by preparation of text, tables, and diagrams may all be involved. Ross Mackay's works clearly demonstrate his competency in all these aspects, but he stands out particularly in his ability to identify problems from field observation and to design and execute field investigations. These characteristics I intend to stress and illustrate, emphasizing his work in that most difficult of terrestrial environments — the Arctic.

NATIVE ATTRIBUTES

Excellent health may not be an absolute necessity for an Arctic scientist but it surely helps. Ross Mackay enjoys the benefits of health. He is almost immune to the biting cold and darkness of the Arctic winter. He can remove his mittens at forty degrees below to write notes or to couple the spade lugs of a probe to its electronic sensor, and has done so without falling prey to frostbite. But it is not just his resistance to cold, but also his precautions — light inner gloves and brief exposure of his hands — that have enabled him to escape this hazard.

Another characteristic of great advantage is a keen memory. To be sure, Dr. Mackay is considered by the Mackenzie Delta eskimos to have a simply atrocious memory and as evidence they point out that he has to write down every little detail in his notebook. My evidence is quite the contrary.

I have walked with him on a bright winter day around the university campus, observing the patterns of stripes created by needle-ice. I have been surprised to find him withdraw from his pocket, first, a compass with which he measured the orientation of these stripes, and then a notebook from a decade earlier in which he had recorded a similar orientation on a particular date at this very same site. He was thus able to produce documentary evidence, not just a recollection by a demonstrably good memory, that the orientation was unchanged.

ACQUIRED CHARACTERISTICS

Preparation

Now lest we have given the impression that Arctic scientists are born rather than made, we should note some acquired or developed characteristics that particularly help Ross Mackay in his work. Probably the foremost of these can be summarized simply as 'preparation.'

He is widely read. At home, much of his time in the winter evenings is devoted to reading: history of the western Canadian Arctic, studies of frozen ground in any polar area, treatises on soil mechanics. With this he becomes armed with questions, ideas, methods, as background for future work.

An example of this type of preparation can be cited from his study of gas-domed pingos. After the unexpected, indeed almost catastrophic, discovery of open space beneath the active layer of a pingo and the discharge of its contained methane (more of this below), he entered the chamber. The frost crystals of the chamber roof caught his attention. The inner layer, in contact with the roof, occurred as plates and needles of ice, shapes which he had learned from his reading could be attributed to crystallization at temperatures of 0°C to -5°C. An outer layer of hollow prismatic columns recorded later temperatures of crystallization between -5°C and -8°C. From this he was able to infer the temperature conditions under which the ice crystals grew and confirm later with temperature measurements that the frost accumulation was the product of a single fall and winter. How many of us would have both the presence of mind and the background needed to observe and record the critical data on these fragile crystal forms before the opportunity was lost?

When he makes his winter visits to the Mackenzie Delta he carries with him a compact repair kit for his electronic gear, complete with a variety of spare parts. Most of the time the latter would prove to be unnecessary, a point that arouses envy and frustration on the part of some colleagues whose equipment does not behave as well as his. When, however, there is a breakdown he almost always has the necessary parts and tools to get his equipment back into service immediately.

Innovation

Included in Mackay's skills is a remarkable capacity for innovation. One of his guiding principles is that "the simplest thing that will do the job is the best thing to do the job." This, incidentally, leads to economical research (more of this later). A recent demonstration of this arose when he found that wooden dowelling, to be used for survey stakes, was becoming hard to obtain and expensive when available. What could he find to substitute for this item? His answer—bamboo chopsticks! These are inexpensive, available in quantity, precut, pre-packed, and are remarkably distinctive when inserted into the Arctic tundra.

A more sophisticated innovation has been spurred by his interest in learning the precise time of day, as well as the time of year, when an ice wedge cracks open with the chilling of permafrost. He had already pioneered the technique of embedding a slender wire in the topsoil, after it had thawed in the summer sun, across the axis of the ice wedge. The wire, later frozen in with the winter cold, would snap at the instant the wedge split open. Accutron clocks are available, at a price, to time the breaking of the wire. Why not, he reasoned, use an inexpensive electronic watch, disconnect one battery terminal and reconnect it via the 'breaking wire.' When the wedge cracks the wire breaks and the watch stops. Moreover a watch that records both the day and the date can be left untended for 7 months before it repeats the pattern of days and dates. As a result it is possible to recapture both the time and the date for the rupture using a single, relatively inexpensive wrist watch.

And how to measure the minimum depth of an ice-wedge crack after it opens? Simply insert a length of stiff surveyor's tape until it can be worked down the crack no farther. And how wide is the crack at depth? Attach to the bottom end of the tape a knob of known diameter. When this jams in the downward-narrowing crack, its width at a measurable depth can be ascertained. An array of such tapes, with a range of terminal knobs, can be used to estimate the rate of taper of the crack with depth.

Curiosity

The happy combination of the curiosity to look and the ability to recognize is another characteristic of Ross Mackay. Details catch both his eye and his interest, whether these be delicate ice needles growing from damp, rotten sticks, the striping of the ground after needle ice has melted, stones frozen within the ice of a pond and lifted off its floor, or the crook developed in the basal trunks of trees by the downslope creep of winter snow.

And not all the features attracting his attention and curiosity need be small. He was the first to associate some topographically high areas along the Beaufort Sea coast with basins directly upstream with respect to the flow of the former ice sheet covering this area. Why, for example, was Herschel Island, standing approximately 10 km square and 150 m high, matched in the waters of Mackenzie Bay by a hollow also about 10 km square and 100 m deep? To satisfy his curiosity he considered the possibility that the ice-thrust beds of Herschel Island had been dragged there by the overriding ice from the hollow to the south-east. When seismic exploration of Mackenzie Bay was later undertaken and the thickness of postglacial beds thereby determined, it was shown that the match in volumes was even better than Mackay had estimated.

Observation and association only whet more curiosity. Why, for example, do ice wedges crack in the winter along the site of cracks of previous years? What makes pingos grow? How are pebbles hoisted into the ice of a freezing pond and which rock types are most susceptible to this process? How fast does a rock stream move? These questions and many more, arising from sharp observation and perceptive recognition of the anomalous or of the unexplained, constitute the curiosity that has served him so well. This curiosity has provided the fuel for his projects.

Dedication

Though his early training allowed him to opt for many careers, Mackay has concentrated on his Arctic exploration. True, he has also looked at alpine and winter conditions farther south, but principally because these were related to the problems of the northern lands and their frozen ground. For more than thirty years now he has devoted a high proportion of his time to this one field of endeavour. He has done much to pioneer winter research in the barren lands of the western Canadian Arctic. During recent years he has devoted up to a week in December and another week in March to the Mackenzie Delta checking on depths of snow, subsurface temperatures, behaviour of ice wedges, changes in dimensions in the patterned ground.

Judgment

Judgment is a quality not easy to assess objectively. It represents a selection amongst choices. What might have followed from a different choice than the one that was made remains a matter of speculation — a 'what if'. Nevertheless, if a high proportion of the choices are the right ones the cumulative effects become recognizable. I suggest, therefore, that the recognition that Mackay has received in his professional career can be credited in part to his exercise of good judgment. Thus, for example, his decision to avoid much of the committee work in favor of research, combined with his decision to set aside time for planning and preparation at the expense of some other activity, plus his choice of instruments or methods to conduct his investigations, and so on, in concert if not individually, contribute to his success as a research worker in his chosen field.

Other judgments lie hidden within the written product of his research. In drawing conclusions he may give the impression of being very conservative. Notwithstanding this impression, he first considers a variety of hypotheses, some of which could indeed be considered outrageous, to explain some phenomenon he is investigating. Those hypotheses that fail to satisfy his evidence he soon rejects, and by degrees he arrives at a very limited number of explanations. He then considers their implications and designs a further test to narrow down the possibilities still more. What he finally commits to print may indeed be conservative but it is not for lack of consideration given to other perhaps less conservative viewpoints. What is presented has been thoroughly judged. The impression of conservatism is created not by the ultimate hypothesis but by his reluctance to speculate in public.

Interpersonal Relations

In his dealings with others during the planning and operation of his studies Ross Mackay is consistently polite, notwithstanding the frustrations all too often experienced these days in the purchase and delivery of material or services. The junior clerk is treated with the same respect as the president of the firm. If the firm fails in its assignment it will, I am sure, be remembered, and where a suitable alternative is available business will be transferred. However, the acquisition of supplies, equipment, or service remains the prime consideration; recrimination made for its own sake plays no part.

This diplomatic approach has paid off. For example, when he sought consent both from native groups and from· government officials for the artificial drainage of a lake near the Arctic coast, no ardent conservationist complained about the forthcoming devastation, nor did officials object to the catastrophic change. He had, of course, already convinced the natives and the authorities that the drainage of this lake by natural coastal retreat was inevitable within a few decades and that the turbidity created would be minor compared with the contribution of the Mackenzie River or of the wave erosion by a periodic storm. And the fact that the drained lake-basin would soon become good goose-hunting ground was not lost on some of the local residents.

Financial Matters

Ross Mackay is particularly circumspect on financial matters. He could have received premium fees for his own services as a consultant to the major companies engaged in the exploration and development of the petroleum resources of the north, both in Alaska and in the Northwest Territories, but he chose not to prostitute his science. He has eschewed all monetary rewards from the companies for the advice he has freely given them. He has accepted non-financial aid for his research, such as logistical support in the field for some of his students, use of company drill holes for some of his own instruments, and company observations on subsurface temperatures and soil types. One of the companies provided him with a field assistant, a trusted employee who, incidentally, learned the details of permafrost while serving on the job. One wonders who benefited most from this happy arrangement.

Financial support for his research has come largely from the Canadian government through the former Geographical Branch, the Geological Survey of Canada, the National Research Council and its successor fund-granting agency, the Natural Sciences and Engineering Research Council of Canada, and the Department of Indian and Northern Affairs. University funds and services have also been a major help. Transportation in the field, a major consideration in Arctic research, has been provided in large measure by the Polar Continental Shelf Project, another federal organization. His policy in authorizing expenditures from research grants is careful and his accounting meticulous. His care in spending coupled with the guiding philosophy that "the simplest thing that will do the job is the best thing to do the job" has given him an enviable reputation among funding organizations in terms of research production per dollar invested.

Views on Transport

Transport, to Ross Mackay, is a means to an end — to reach the site for his investigations. Walking and snowshoeing are legitimate if slow and inefficient. Skiing is, surprisingly, not one of his favored means of locomotion. The appeal of the snowmobile is also limited. Neither mode of travel appears to be completely controlled and safe.

In his early summer operations he first adopted the boat as a means of reaching sites along both Mackenzie River and the Arctic coast. For a few years he was accustomed to ordering a freight canoe to be delivered to Fort Providence, near the outlet of Great Slave Lake, from which point he would use it to cruise down Mackenzie River and work in the Delta for the summer. He would sell the canoe at Inuvik at the end of the season and fly out. He was introduced to the helicopter rather belatedly, and this has proved to be a persisting love affair. He is now convinced that no Arctic explorer should be without one.

Views on Adventure

Adventure is something a good scientist shuns. To Ross Mackay an adventure, in which something unexpected takes place, represents some miscalculation. Even such serendipitous adventure as the discovery of methane supporting the 'gas domed pingos' he investigated in 1963 falls into this category. The drilling of a small mound had been planned and a suitable power drill acquired and transported to the site. One of several pingos was chosen for investigation and a hole was started into the summit of the mound. Suddenly the drill stem and motor dropped to the ground as the bit entered open space. With the withdrawal of the drill gas started escaping from the hole. What gas? A match was struck to test the gas and it became immediately obvious that the gas was flammable. Soon the tundra adjacent to the pillar of burning gas was alight and with it the residue of gasoline on the keg of fuel. Some minutes were required to extinguish the fires. Notwithstanding the happy ending of this adventure Mackay, I am sure, wished, most particularly during those few tense minutes, for a better planned investigation.

Another adventure occurred while he was captain of the *Tulik*, a thirty foot schooner used for his early exploration of the Beaufort Sea coast. During one of his sessions at the helm, one of his two crewmen staggered from below decks in an obviously confused condition. Diagnosing the problem as a case of carbon monoxide poisoning, he had to rescue the second assistant from below and lash both of the dazed victims to the mast, all the while keeping watch that the vessel was not heading for trouble. Once again the adventure had no lasting ill effects; once again Mackay wished it had never happened. Still another adventure was forestalled by his practice of checking his instruments. We were camped on the treeless coastal lowlands of northernmost Yukon Territory studying the local geomorphology to better understand the geological history of an archeological site. On one of the first days the camp was thoroughly and persistently fogbound. I undertook to make some measurements in camp, but Mackay wanted to pursue an investigation well out in the field and left our tent to set out by himself into the wilderness. Within minutes he was back in the tent, thrusting his pocket compass into my hands with the request that I check it. I soon identified his problem—the north end of the compass needle was pointing south! Had he slipped off into the fog before checking his bearing he would likely have become disoriented and might well have strayed far from camp before breaking clouds could reveal the compass error—an adventure averted this time!

Pure and Applied Research

All too often these days there are attempts to identify some scientific endeavour as either 'pure' or 'applied,' the former being motivated by curiosity and its results, by implication, of no immediate practical value to mankind; and the latter being of practical value, perhaps being of direct financial benefit, even though it may not advance the frontiers of science. Mackay's career clearly demonstrates the weakness of this classification. His work throughout his career has been undertaken to satisfy a very active curiosity and with no claim for financial benefit or personal gain. For some eighteen years his efforts would have been designated 'pure science' until the discovery of oil at Prudhoe Bay. Then a very urgent need arose to understand the behaviour of ice-rich permafrost penetrated or overlain by pipes carrying hot oil. Mackay's pioneer work suddenly became transformed into very practical 'applied science.' Nothing had changed in the nature of his work or of his motivation; all that had altered was someone else's conception of its value.

THE CONTRIBUTION OF ROSS MACKAY

Ross Mackay's published work represents a model of the application of scientific method in field science. Each paper defines a particular problem in a manner that permits examination, outlines the methods adopted to probe it, presents the evidence gathered in the field, considers alternative resolutions, applies the evidence in order to eliminate all but the most plausible explanation and then states the conclusion and its implications. He has developed properly controlled experiments to reveal the nature of various permafrost phenomena. This is a rare feat in a field science. Nevertheless, it is the little noted but essential attributes of character discussed in this chapter that have permitted Mackay's consistently successful application of method in the field.

The fruits of his work appear in over 160 papers and monographs (to date), mostly single-authored and every one containing some new idea. With most of his scientific enquiries directed toward northern lands and their physical properties, it is easy to overlook some earlier statements in the fields of cartography, regional analysis and 'social physics' which are fundamental and still quoted. The bibliography of Ross Mackay's writings reveals both the range of his contributions and the enormous depth of his work in geocryology.

PUBLICATIONS BY J. ROSS MACKAY

The north shore of the Ottawa river. Rev. Can. de Géographie *1* (2-3), 1947: 3-8.

Dotting the dot map: an analysis of dot size, number and visual tone density. Surveying and Mapping *9*, 1949: 3-10.

Physiography of the lower Ottawa valley. Rev. Can. de Géographie *3*, 1949: 53-96.

Some problems and techniques in isopleth mapping. Ec. Geography *27*, 1951: 1-9. Reprinted in Surveying and Mapping *12*, 1952: 14-18.

Physiography of the Darnley Bay area, N.W.T. Can. Geographer No. 2, 1952: 31-34.

A new projection for cubic symbols on economic maps. Ec. Geography *29*, 1953: 60-62.

Percentage dot maps. Ec. Geography *29*, 1953: 263-266.

The alternative choice in isopleth interpolation. Prof. Geographer *5* (4), 1953: 2-4.

The atlas habit. The B.C. Teacher *32*, 1953: 252-254.

Fissures and mud circles on Cornwallis Island, N.W.T. Can. Geographer No. 3, 1953: 31-37.

Post-glacial drainage changes in the Darnley Bay area, N.W.T., Canada. Assoc. Pacific Coast Geographers, Yrbk. *15*, 1953: 17-22.

Arithmetic-square root graph paper. Prof. Geographer *6* (1), 1954: 15-16.

Geographic cartography. Can. Geographer No. 4, 1954: 1-14.

Notes on the planetary-wind diagram. J. Geography *53*, 1954: 154-156.

An analysis of isopleth and choropleth class intervals. Ec. Geography *31*, 1955: 71-81.

Physiography of the Northlands. *In* G.H.T. Kimble and D. Good, editors, Geography of the Northlands. Amer. Geog. Soc. & John Wiley & Sons, 1955: Ch. 2, 11-35.

Percentage isopleth maps. Prof. Geographer *7* (6), 1955: 10-12.

Physical and biogeography in Canada (mid-1952 to mid-1954). Can. Geographer No. 6, 1955: 21-23.

Mackenzie deltas — a progress report. Can. Geographer No. 7, 1956: 1-12.

Surficial geology of the Firth river archaeological site, Yukon Territory. Arctic *9*, 1956: 210-211 (with W.H. Mathews).

Deformation by glacier-ice at Nicholson Peninsula, N.W.T., Canada. Arctic *9*, 1956: 218-228.

Notes on oriented lakes of the Liverpool Bay area, N.W.T. Rev. Can. de Géographie *10*, 1956: 169-173. Discussion, *11*, 1957: 175-178.

Field observation of patterned ground. Can. Alpine J. *40*, 1957: 91-96.

Notes on small boat harbours, N.W.T. Canada Dept. Mines and Technical Surveys, Geog. Br. Geog. Paper *13*, 1957: 11 pp.

Chi square as a tool for regional studies. Assoc. Amer. Geographers, Ann. *48*, 1958: 164.

Arctic 'vegetation arcs.' Geog. J. *124*, 1958: 294-295.

The interactance hypothesis and boundaries in Canada: a preliminary study. Can. Geographer No. 11, 1958: 1-8.

Conformality: mathematical and visual. Prof. Geographer *5*, 1958: 12-13.

A subsurface organic layer associated with permafrost in the western Arctic. Canada Dept. Mines and Technical Surveys, Geog. Br. Geog. Paper 18, 1958: 21 pp.

The valley of the lower Anderson River, N.W.T. Geog. Bull. No. 11, 1958: 37-56.

The Anderson River map area, N.W.T. Canada Dept. Mines and Technical Surveys, Geog. Br. Memoir *5*, 1958: 137 pp.

Mackenzie delta coastland. Assoc. Pacific Coast Geographers, Yrbk. *20*, 1958: 53-54.

Appendix in Hohn, E.O., Birds of the Anderson River. Can. Field Naturalist *73*, 1959: 113 (93-114).

Comments on the use of chi-square. Assoc. Amer. Geographers Ann. *49*, 1959: 89.

Regional geography: a quantitative approach. *In* Mélanges géographiques canadiens offerts à Raoul Blanchard. Les presses de l'Université Laval, 1959: 57-63.

Glacier ice-thrust features of the Yukon coast. Geog. Bull. No. 13, 1959: 5-21.

Crevasse fillings and ablation slide moraines, Stopover Lake area, N.W.T. Geog. Bull. No. 14, 1960: 89-99.

Deformation of soils by glacier ice and the influence of pore pressure and permafrost. Roy. Soc. Canada, Trans. Sec. IV, 3rd Ser., *54*, 1960: 27-36 (with W.H. Mathews).

Notes on small boat harbours of the Yukon coast. Geog. Bull. No. 15, 1960: 19-30.

Geology of the Engigsteiak Archaeological Site, Yukon Territory. Arctic *14*, 1961: 25-52. (with W.H. Mathews and R.S. MacNeish).

A study of freeze-up and break-up at Fort Good Hope, N.W.T. *In* Thought. Toronto, W.J. Gage, 1961: 65-71.

Geomorphology and the geography student. Can. Geographer, *5* (3), 1961: 30-33.

Freeze-up and break-up of the lower Mackenzie river, N.W.T. *In* Raasch, G.O., Geology of the Arctic. Univ. Toronto Press, 1961: 1119-1134.

Pingos of the Pleistocene Mackenzie delta area. Geog. Bull. No. 18, 1962: 20-63.

Some cartographical problems in the field of special (thematic) maps. Can. Cartography *1*, 1962: 42-47.

Origin of the pingos of the Pleistocene Mackenzie delta area. 1st Canadian Conf. on Permafrost, Proc. NRCC Tech. Mem. *76*, 1963: 79-82.

Progress of break-up and freeze-up along the Mackenzie river. Geog. Bull. No. 19, 1963: 103-116.

Notes on the shoreline recession along the coast of the Yukon Territory. Arctic *16*, 1963: 195-197.

Snowcreep studies, Mount Seymour, B.C.: preliminary field investigations. Geog. Bull. No. 20, 1963: 58-75. (with W.H. Mathews).

Isopleth class intervals: a consideration in their selection. Can. Geographer *7*, 1963: 42-45.

The Mackenzie delta area, N.W.T. Canada Dept. Mines and Technical Surveys, Geog. Br., Mem. *8*, 1963: 202 pp.

Pollen diagrams in the Mackenzie delta area, N.W.T. Arctic *16*, 1963: 229-238 (with J. Terasmae).

Arctic Landforms. *In* The unbelievable lands. I.N. Smith, ed. Ottawa, The Queen's Printer, 1964: 60-62.

The role of permafrost in ice-thrusting. J. Geol. 72, 1964: 378-380. (with W.H. Mathews). Discussion, v. *73*, 1965: 896.

Glacier flow and analogue simulation. Geog. Bull. *7*, 1965: 1-6.

Historical records of freeze-up and break-up on Churchill and Hayes rivers. Geog. Bull. *7*, 1965: 7-16 (with D.K. Mackay).

Gas-domed mounds in permafrost, Kendall Island, N.W.T. Geog. Bull. *7*, 1965: 105-115.

Segregated epigenetic ice and slumps in permafrost, Mackenzie Delta area, N.W.T. Geog. Bull, *8*, 1966: 59-80.

Thick tilted beds of segregated ice, Mackenzie delta area, N.W.T. Biul. Peryglacjalny *15*, 1966: 39-43.

Pingos in Canada. Permafrost Int. Conf., 1963, Proc. NAS-NRC (U.S.A.), Pub. *1287*, 1966: 71-76.

The mixing of the waters of the Liard and Great Bear rivers with those of the Mackenzie. Geog. Bull. *8*, 1966: 166-173.

Tundra and taiga. *In* F.F. Darling and J.P. Milton, eds., Future Environments of North America. Garden City, N.Y., Natural History Press, 1966: 156-171.

The structure of some pingos in the Mackenzie delta area, N.W.T. Geog. Bull. *8*, 1966: 360-368 (with J.K. Stager).

Needle ice and seedling establishment in southwestern B.C. Can. J. Plant Science *47*, 1967: 135-139 (with V.C. Brink, S. Freyman, and D.G. Pearce).

Permafrost depths, lower Mackenzie valley, Northwest Territories. Arctic *20*, 1967: 21-26.

Underwater patterned ground in artificially drained lakes, Garry Island, N.W.T. Geog. Bull. *9*, 1967: 33-44.

Freeze-up and break-up prediction of the Mackenzie River, N.W.T., Canada. *In* Quantitative Geography, Part II. Northwestern University, Evanston, Illinois, Department of Geography, 1967: 25-66.

Physiography. Naval Arctic Manual ATP 17 (A), Part I. Environment. Arctic Inst. North America, Montreal, 1967: Ch. 1, 1-27 (with J.B. Bird and J.P. Johnson).

Observations on pressures exerted by creeping snow, Mount Seymour, British Columbia, Canada. Int. Conf. of Low Temp. Science, Inst. of Low Temp. Science, Hokkaido University, Japan, 1967. Physics of Snow and Ice *I*, Pt. 2: 1185-1197.

Discussion of the theory of pingo formation by water expulsion in a region affected by subsidence. J. Glaciology 7 (50), 1968: 346-350.

The Mackenzie delta. Can. Geog. J. 57 (5), 1969: 146-155.

The perception of conformality of some map projections. Geog. Rev. 59, 1969: 373-387.

Lateral mixing of the Liard and Mackenzie rivers downstream from their confluence. Can. J. Earth Sciences 7, 1970: 111-124.

Geomorphic processes, Mackenzie valley, Arctic coast, District of Mackenzie. Geol. Surv. Canada, Rep. Activities, 1969, Part A: 197-198.

Disturbances to the tundra and forest tundra environment of the Western Arctic. Can. Geotech. J. 7, 1970: 420-432.

Geomorphic processes, Mackenzie valley, Arctic coast, District of Mackenzie. Geol. Surv. Canada, Rep. Activities, 1970, Part A: 189-190 (1971).

Massive ice and icy sediments throughout the Tuktoyaktuk Peninsula, Richards Island, and nearby areas, District of Mackenzie. Geol. Surv. Canada Paper 71-21, 1971: 16 pp. (with V.N. Rampton).

The origin of massive icy beds in permafrost, western Arctic coast, Canada. Can. J. Earth Sciences 8, 1971: 397-422.

Application of $H^{18}O/H_2^{16}O$ abundances to the problem of lateral mixing in the Liard-Mackenzie river system. Can. J. Earth Sciences 8, 1971: 1107-1109 (with H.R. Krouse).

Contributor to Blais, R.A. and Smith, C.H., editors, Earth sciences serving the nation. Science Council of Canada, Special Study 13, 1971: 363 pp.

Geomorphic processes, Mackenzie valley, Arctic coast, District of Mackenzie. Geol. Surv. Canada Paper 72-1A, 1972: 192-194.

Ground ice in the active layer and the top portion of permafrost. Permafrost Seminar on the Active Layer, Proc. NRCC, Tech. Mem. 103, 1971: 26-30.

The world of underground ice. Assoc. Amer. Geog. Ann. 62, 1972: 1-22.

Some observations on the growth of pingos. In D.E. Kerfoot, ed., Mackenzie delta area monograph. Brock University, St. Catharines, Ontario, for 22nd Int. Geog. Cong. 1972: 141-147.

Some observations on ice-wedges, Garry Island, N.W.T., ibid., 131-139.

Ground temperatures at Garry Island, N.W.T. ibid. 107-114 (with D.K. Mackay).

Break-up and ice-jamming of the Mackenzie River, Northwest Territories, ibid. 87-93 (with D.K. Mackay).

Geomorphological process studies, Garry Island, N.W.T., ibid., 115-130 (with D.E. Kerfoot).

Permafrost and ground ice. Can. Northern Pipeline Research Conf. Proc., NRCC, Tech. Mem. 104, 1972: 235-239.

Relic Pleistocene permafrost, western Arctic, Canada. Science 176, 1972: 1321-1323 (with V.N. Rampton and J.G. Fyles).

Application of water temperatures to the problem of lateral mixing in the Great Bear-Mackenzie river system. Can. J. Earth Sciences 9, 1972: 913-917.

Recent and late Pleistocene permafrost in the western Arctic of Canada. 21st Int. Geog. Cong., New Delhi, India, Selected Papers 1, 1970: 72-74.

Offshore permafrost and ground ice, southern Beaufort Sea, Canada. Can. J. Earth Sciences *9*, 1972: 1550-1561.

Some aspects of permafrost growth, Mackenzie delta area, N.W.T. Geol. Surv. Canada Paper *73-1A*, 1973: 232-233.

Geomorphology and Quaternary history of the Mackenzie river valley, near Fort Good Hope, N.W.T., Canada. Can. J. Earth Sciences *10*, 1973: 26-41 (with W.H. Mathews).

Origin, composition, and structure of perenially frozen ground and ground ice. Permafrost 2nd Int. Conf. Yakutsk, U.S.S.R. 13-28 July, 1973: North American Contribution. Washington, D.C., NAS (USA), 1973: 185-192 (with R.F. Black).

Massive ground ice, western Arctic, Canada. Permafrost 2nd Int. Conf. Yakutsk, U.S.S.R., 13-28 July, 1973: North American Contribution. Washington, D.C., NAS (USA), 1973: 223-228.

Winter cracking (1967-1973) of ice-wedges, Garry Island, N.W.T. Geol. Surv. Canada Paper *73-1B*, 1973: 161-163.

The growth of pingos, western Arctic coast, Canada. Can. J. Earth Sciences, *10*, 1973: 979-1004.

A frost tube for the determination of freezing in the active layer above permafrost. Can. Geotech. J. *10*, 1973: 392-396.

Break-up and ice jamming on the Mackenzie River, N.W.T. *In* Hydrologic aspects of northern pipeline development. Canada, Environmental-Social Committee, Northern Pipelines, Task Force on Northern Oil Devpt., Rep. *73-3*, 1973: 223-232 (with D.K. Mackay).

Locations of spring ice jamming on the Mackenzie river, N.W.T. *In* Hydrologic aspects of northern pipeline development. Canada, Environmental-Social Committee, Northern Pipelines, Task Force on Northern Oil Devpt., Rep. *73-3*, 1973: 233-257 (with D.K. Mackay).

The rapidity of tundra polygon growth and destruction, Tuktoyaktuk Peninsula-Richards Island area, N.W.T. Geol. Surv. Canada Paper, *74-1A*, 1974: 391-392.

Seismic shot holes and ground temperatures, Mackenzie delta area, N.W.T. Geol. Surv. Canada Paper *74-1A*, 1974: 389-390.

Measurement of upward freezing above permafrost with a self-positioning thermistor probe. Geol. Surv. Canada Paper *74-1B*, 1974: 250-251.

Ionic and oxygen isotopic fractionation in permafrost growth. Geol. Surv. Canada *74-1B*, 1974: 255-256 (with L.M. Lavkulich).

Performance of a heat transfer device, Garry Island, N.W.T. Geol. Surv. Canada Paper *74-1B*, 1974: 252-254.

Reticulate ice veins in permafrost, northern Canada. Can. Geotech. J. *11*, 1974: 230-237. Discussion v. *12*, 1975: 163-165.

Needle ice striped ground. Arctic and Alpine Research *6*, 1974: 79-84 (with W.H. Mathews).

Movement of sorted stripes, The Cinder Cone, Garibaldi Park, B.C., Canada. Arctic and Alpine Research *6*, 1974: 347-359 (with W.H. Mathews).

Ice-wedge cracks, Garry Island, N.W.T. Can. J. Earth Sciences *11*, 1974: 1366-1383.

Relict ice wedges, Pelly Island, N.W.T. Geol. Surv. Canada Paper *75-1A*, 1975: 469-470.

Freezing processes at the bottom of permafrost, Tuktoyaktuk Peninsula, District of Mackenzie. Geol. Surv. Canada Paper *75-1A*, 1975: 471-474.

Snow creep: its engineering problems and some techniques and results of its investigation. Can. Geotech. J. *12*, 1974: 187-198 (with W.H. Mathews).

Heat energy of the Mackenzie River. *In* Further hydrologic studies of the Mackenzie valley. Canada, Environmental-Social Committee, Northern Pipelines, Task Force on Northern Oil Devpt., Rep. *74-35*, 1975: 1-23 (with D.K. Mackay).

The closing of ice-wedge cracks in permafrost, Garry Island, N.W.T. Can. J. Earth. Sciences *12*, 1975: 1668-1674.

The stability of permafrost and recent climatic change in the Mackenzie valley, N.W.T. Geol. Surv. Canada Paper *75-1B*, 1975: 173-176.

Some resistivity surveys of permafrost thickness, Tuktoyaktuk Peninsula, N.W.T. Geol. Surv. Canada Paper *75-1B*, 1975: 177-180.

Lake bottom freezing and periglacial effects in western Canadian Arctic. *In* 21st Int. Geog. Cong., Selected Papers *4*: Regional Geography and Cartography. Can. National Committee Geography, 1972: 1.

Pleistocene permafrost, Hooper island, Northwest Territories. Geol. Surv. Canada Paper *76-1A*, 1976: 17-18.

Ice-wedges as indicators of recent climatic change, western Arctic coast. Geol. Surv. Canada Paper *76-1A*, 1976: 233-234.

Ice segregation at depth in permafrost. Geol. Surv. Canada Paper *76-1A*, 1976: 287-288.

The age of Ibyuk Pingo, Tuktoyaktuk Peninsula, District of Mackenzie. Geol. Surv. Canada Paper *76-1B*, 1976: 59-60.

On the origin of pingos — a comment. J. Hydrology *30*, 1976: 295-298.

The growth of ice wedges (1966-1975), Garry Island, N.W.T., Canada. 23rd Int. Geog. Congress, Moscow, Proc. Geomorphology and Paleogeography, 1976: 180-182.

Cryostatic pressures in nonsorted circles (mud hummocks), Inuvik, Northwest Territories. Can. J. Earth Sciences *13*, 1976: 889-897 (with D.K. Mackay).

Probing for the bottom of the active layer. Geol. Surv. Canada Paper *77-1A*, 1977: 327-328.

The widths of ice wedges. Geol. Surv. Canada Paper *77-1A*, 1977: 43-44.

Permafrost growth and subpermafrost pore water expulsion, Tuktoyaktuk Peninsula, District of Mackenzie. Geol. Surv. Canada Paper *77-1A*, 1977: 323-326.

Pulsating pingos, Tuktoyaktuk Peninsula, N.W.T. Can. J. Earth Sciences, *14*, 1977: 209-222.

Changes in the active layer from 1968 to 1976 as a result of the Inuvik fire. Geol. Surv. Canada Paper *77-1B*, 1977: 273-275.

Reliability of permafrost thickness determination by D.C. resistivity soundings. NRCC Tech. Mem. *119*, 1977: 25-39 (with W.J. Scott).

The stability of ice-push features, Mackenzie River, Canada. Can. J. Earth Sciences, *14*, 1977; 2213-2225 (with D.K. Mackay).

Needle ice growth, ice segregation, and frost heave under natural conditions. *In* Energy, water, and the physical environment of the soil. B.C. Ministry of Agriculture, Victoria. 6th B.C. Soil Science Workshop Rep., 1977: 173-178.

Sub-pingo water lenses, Tuktoyaktuk Peninsula, Northwest Territories. Can. J. Earth Sciences, *15*, 1978: 1219-1227.

The surface temperature of an ice-rich melting permafrost exposure, Garry Island, Northwest Territories. Geol. Surv. Canada Paper *78-1A*, 1978: 551-552.

The use of snow fences to reduce ice-wedge cracking, Garry Island, Northwest Territories. Geol. Surv. Canada Paper *78-1A*, 1978: 523-524.

Freshwater shelled invertebrate indicators of paleoclimate in northwestern Canada during late glacial times: Discussion. Can. J. Earth Sciences, *15*, 1978: 461-462.

Contemporary pingos. Biul. Periglacjalny *27*, 1978: 133-154.

Quaternary and permafrost features, Mackenzie delta area. *In* F.G. Young, ed., Geological and geographical guide to the Mackenzie delta area, Can. Soc. Petroleum Geologists, 1978: 42-50.

An equilibrium model for hummocks (non-sorted circles), Garry Island, Northwest Territories. Geol. Surv. Canada Paper *79-1A*, 1979: 165-167.

Frost heave at below 0°C ground temperature, Inuvik, Northwest Territories. Geol. Surv. Canada Paper *79-1A*, 1979: 403-405 (with J. Ostrick, C.P. Lewis, and D.K. Mackay).

Uplift of objects by an upfreezing ice surface. Can. Geotech. J. *16*, 1979: 609-613 (with C. Burrous).

Pingos of the Tuktoyaktuk Peninsula area, Northwest Territories. Géographie Phys. Quat. *23*, 1979: 3-61.

Geologic controls of the origin, characteristics, and distribution of ground ice. In 3rd Int. Conf. on Permafrost, Edmonton, Canada, 10-13 July, 1978. Proc. *2*. Ottawa, NRC Canada Pub. 16529, 1979: 1-18 (with V.N. Konishchev and A.I. Popov).

Ridge growth of ice-wedge polygons, western Arctic, Canada. 24th Int. Geog. Cong., Absts *1*, 1980: 86-87.

Deformation of ice-wedge polygons, Garry Island, Northwest Territories. Geol. Surv. Canada Paper *80-1A*, 1980: 287-291.

The origin of hummocks, western Arctic coast, Canada. Can. J. Earth Sciences *17*, 1980: 996-1006.

An experiment in lake drainage, Richards Island, Northwest Territories: a progress report. Geol. Surv. Canada Paper *81-1A*, 1981: 63-68.

Frost heave at Inuvik. 4th Can. Permafrost Conf., Calgary, Alberta 1981: 116-117 (with C.P. Lewis).

Dating the Horton river breakthrough, District of Mackenzie. *In* Current Research, Part B. Geol. Surv. Canada Paper *81-1B*, 1981: 129-132.

Active layer slope movement in a continuous permafrost environment, Garry Island, Northwest Territories, Canada. Can. J. Earth Sciences *18*, 1981: 1666-1680.

Aklisuktuk (Growing Fast) Pingo, Tuktoyaktuk Peninsula, Northwest Territories, Canada. Arctic *34*, 1981: 270-273.

2
EXPERIMENTAL OBSERVATIONS OF PERIGLACIAL PROCESSES IN THE ARCTIC

Alfred Jahn

Geographical Institute, University of Wrocław
pl. Uniwersytecki, 1, 50-137 Wrocław Poland

ABSTRACT

The author describes the beginnings of experimental field observations of periglacial processes, which he started in 1937 in Greenland. The observations concerned mainly the active layer over permafrost and led to empirical formulation of the rule of summer thawing of frozen ground according to the square root of time after the commencement of thaw. Research was continued in Spitsbergen from 1957 to 1978. As a result of these studies the relation between the active layer depth and slope inclination was recognized. The author discusses also the method and results of experimental observations on the problems of the frost movement of soil, gelifluction and water action in the periglacial environment.

RÉSUMÉ:
Observations expérimentales des processus périglaciaires dans l'Arctique

L'auteur décrit le début de ses observations expérimentales de terrain sur les processus périglaciaires, commencées au Groenland en 1937. Les observations portaient surtout sur la couche active au dessus du pergélisol et aboutirent à la formulation empirique d'une loi de fonte du gélisol en été selon la racine carrée du temps après le début du dégel. Les recherches se poursuivirent au Spitsbergen durant les années 1957-1978. Là, fut reconnut le rapport entre la profondeur de la couche active et l'inclinaison de la pente. L'auteur discute également de la méthode et des résultats d'observations expérimentales sur les problèmes de solifluxion, de gélifluxion, et de l'action des eaux courantes dans le milieu périglaciaire.

ЭКСПЕРИМЕНТАЛЬНЫЕ НАБЮДЕНИЯ НАД ПЕРИГЛЯЦИАЛЬНЫМИ ПРОЦЕССАМИ В АРКТИКЕ

А. ДЖАН

РЕЗЮМЕ

Автор описывает начальные стадии экспериментальных полевых наблюдений перигляцияльных процессов, которые он начал в 1937 г в Гринландии. Наблюдения относились главным образом к деятельному слою, перекрывающему многолетнюю мерзлоту и привели к эмпирическому сформулированию закономерности летнего протаивания многолетней мерзлоты согласно квадратному корню времени после начала протаивания. Исследования продолжались в период от 1957 г до 1978 г на островах Шпицберген. Была обнаружена зависимость между глубиной деятельного слоя и углом склона. Автор тоже обсуждает методику и результаты экспериментальных наблюдений, относящихся к проблеме продвижения мерзлоты в породе, олединения (джелифлекции) и действия воды при перигляцияльных условиях.

SOME REMARKS CONCERNING THE HISTORY OF PERIGLACIAL RESEARCH.

This volume is dedicated to Professor J. Ross Mackay who has achieved outstanding results in periglacial research using the method of field experiments. His main field laboratory, on Garry Island in the Mackenzie Delta, is something unique in the world. The objectives of his research are to study processes directly related to permafrost and the active layer and the problem of ground ice. The advantage of his experiments lies in the fact that they are not limited to immediate, short-term effects, but include long-term processes working in the natural periglacial environment of the tundra. He makes use of commercially available equipment, usually modified by himself in accordance with specific experimental needs. For comparison it is worthwhile to present some of the experimental observations in which I have been involved for more than 40 years, in the context of a brief history of periglacial research.

In every science there are problems and subjects which, having been for a time the subject of mere conjectures and suppositions, suddenly gain importance and arouse tremendous interest, or again sink imperceptibly into oblivion. In geography and geology a typical example of such vicissitudes is provided by the history of periglacial problems. The first period of activity falls about the time of the 11th Geological Congress in Stockholm, in 1910, after which some of the participants went to Spitsbergen in order to become acquainted with contemporary periglacial phenomena. This was a time of lively discussions on the genesis of "structured soil," block fields and solifluction. The second period to witness an outburst of interest in periglacial problems dates back to the years which immediately preceded the outbreak of the Second World War; this period still continues. The scope of problems has widened greatly in this period and the main current of thought has been directed towards the Pleistocene epoch.

Both these periods of increased activity in the field of periglacial study saw seminal publications by Walery Lozinski, Professor at the Jagellonian University in Cracow (Jahn, 1954). In 1909 his treatise, "On the mechanical weathering of sandstones in temperate climates" appeared as a publication of the Polish Academy of Sciences and Letters. In this paper Lozinski set forth a new concept which he termed "periglacjalna facya" (periglacial facies). At the time of its appearance, only three years had passed since the publication of the paper on solifluction by Andersson (1906), in which the author had formulated the first definition of periglacial climate and periglacial facies. Another event of particular importance for the development of periglacial research was the appearance of the work by Högbom (1914) entitled "On the geological significance of frost." Just at that time Meinardus (1912) introduced a notion of the so-called "structured soils" (Strukturboden), called in English "patterned ground" (Washburn, 1956) and known under that name subsequently to modern science. Both Meinardus and his followers tried to give a precise description of these exceptionally regular forms on the surface of the arctic soil; they analyzed the sub-surface structure of these soils and put forward hypotheses concerning their origin. This period in the history of periglacial studies resulted in a monograph written by Carl Troll (1944) which is considered to be the most complete study of periglacial phenomena and summarizes the research accomplished before the Second World War.

Along with the research which clearly defined its focus as periglacial phenomena and land forms, there developed a science of permafrost, first of all in Siberia. It is worth adding that the Polish engineer, Leonard Jaczewski, working on the construction of the trans-Siberian railway, was the first to formulate a research problem of "vecna merzlota" in his paper published in the Journal of the Russian Geographic Society in 1889. For the Russians, permafrost was first of all a technological problem connected with building operations in Siberia. Their longstanding research before the Second World War resulted in a fine paper written by the leading expert in this field (M.J. Sumgin, 1937). At the same time there were investigations of permafrost carried out in Alaska and northern Canada. The highest achievement here was the work by Leffingwell (1919).

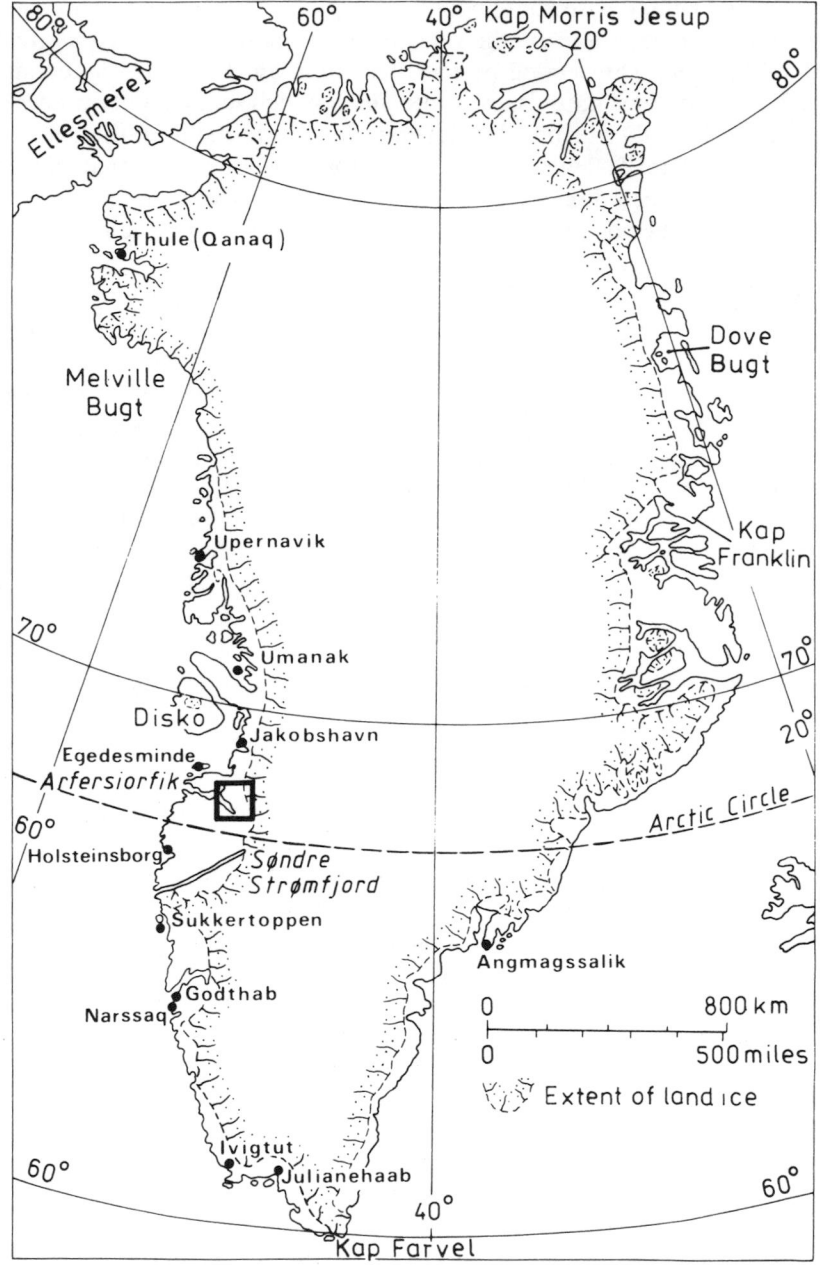

Fig. 2-1. Map of Greenland. The area investigated by the Polish research expedition in 1937 is denoted by a square.

Looking back at the long history of periglacial studies before the Second World War we can see that they were mainly of a descriptive character. At that time there was no experimental, field or laboratory research. There were some limited attempts at measuring the dynamics (rate) of frost processes but they were not very successful.

The author of this paper was amongst the first experimenters: in 1937, as a member of the Polish expedition to Greenland led by A. Kosiba, I carried out experimental observations in the tundra of the Arfersiorfik Fjord

Fig. 2-2. The camp of the Polish research expedition in Arfersiorfik Fiord, West Greenland, just at the edge of inland-ice. In the foreground, near the tents, the so-called tundra craters, i.e., loam boils in the active layer, July 1937.

(figure 2-1). We pitched our camp in front of the inland-ice on a low coastal terrace (figure 2-2). My observations lasted throughout the summer, from May to September, 1937, and the results were published immediately after the war (Jahn, 1946) in Polish.

The observations concerned the following problems: (1) formation of the seasonally thawing ground (active layer); (2) movement of the soil in the tundra environment; and (3) displacement of blocks on slopes ("ploughing blocks"). Very simple instruments were used, such as a steel rod for measuring the depth of soil thawing, and a set of ground thermometers (figure 2-3). Block movement was measured by inserted bench-marks and a steel tape.

Fig. 2-3. Measurement of the thickness of the active layer and of soil temperature in Arfersiorfik Fiord, West Greenland. The photograph shows the author, July 1937.

Immediately after the war I could not continue my work in Greenland. My next opportunity arose for periglacial field work in 1957, when a Polish Research Station was organized in Hornsund Fjord in Spitsbergen (figure 2-4). Being equipped with better instruments I could then measure and evaluate the rates of some periglacial processes.

THE DEPTH OF SUMMER THAWING OF THE GROUND.

Summer thawing in Greenland began at the end of May. The sounding of the ground and the measurement of soil temperature indicated that the depth of thawing in June was on average 26 cm. The increase in the thickness of the thawed layer in July was 13 cm, and in August only 6 cm. On the basis of these data, in the report from the expedition (Jahn, 1946) I formulated a general conclusion — for the first time in periglacial research — which appeared as a kind of law. The increase in thickness of the active layer is approximately proportional to the square root of time:
$$Z_{th} = \alpha \sqrt{t}$$
where Z_{th} denotes the depth of the thaw plane, t is elapsed time from the

Fig. 2-4. Map of islands in the Svalbard archipelago showing locations of Polish periglacial research activities.

commencement of thawing, and α is a constant. Of course, we recognize this equation today as a special form of the general equation for conductive heat transfer, elaborated for freezing and thawing ground by Terzaghi (1952).

My later studies on Spitsbergen confirmed the law. I made observations using the same method as in Greenland, that is, with a thin steel sounding rod. I measured the thickness of the thawing ground every day, or at some other time intervals, on a low tundra in Hornsund Fjord. In the first days thawing was rather quick, but soon it slowed down because of the growing thickness of the upper insulating layer of the soil. The measurements were conducted under various surface conditions, covered with or devoid of vegetation. All the curves (figure 2-5a) show the same trend. Czeppe (1961)

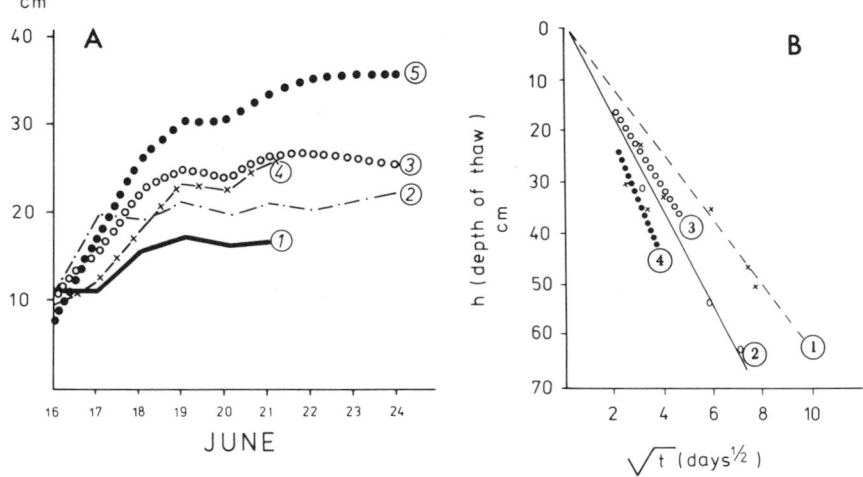

Fig. 2-5. (a) Ground thawing on the coastal terrace near the Polish Station in Hornsund, just after melting of winter snow in June 1974; (1) (2) plots with unchanged natural conditions, covered with vegetation; (3) a plot with the vegetation cover removed; (4) (5) plots with the thawed layer removed on the day of the measurements.
(b) Summer thawing of soil in Greenland and Spitsbergen; (1) Greenland, 1937; (2) Spitsbergen, 1958 (Z. Czeppe, 1961; A. Jahn, 1961); (3) Spitsbergen, 1974 with vegetation cover; (4) Spitsbergen, 1974 with no vegetation cover. Note the inversion of the ordinate axis between the two diagrams.

came to the same conclusion after his observations in Hornsund in the summer of 1958.

The above-mentioned pattern of summer changes of the frost table depends upon one parameter, the amount of heat penetrating into the soil. That is related to what Ryden and Kostov (1980) later called a "thawing index" determined as "a sum of daily mean air temperatures above zero." It is obvious that the active layer is controlled by many factors, as for example, summer and autumn rainfalls which saturate soil with water and change its thermal conductivity. Thus, it may happen that even at the end of the summer season, after a rapid rainfall, the deep thawing process can be rejuvenated and accelerated for several days. However, the general decrease of thawing processes in the first half of summer is the rule. The results are confirmed by observations in Alaska (Brown, Rickard and Victor, 1964; Walker and Harris, 1976; Bilgin, in Brown, Haugen and Parrish, 1975). Summer thawing of soil usually ends in August and at the end of summer the layer becomes stabilized.

The lower part of the active layer has a marked concentration of mineral salts, since ground water is retained there for a longer time during the second half of summer. If this situation recurs periodically for many years — as it does in Siberia — the outcrops of the active layer at river banks have a characteristic yellow-brown strip marking the oxidation zone of the ground water retained on the permafrost substratum.

The results of measuring the depth of soil thawing in Greenland and Spitsbergen are plotted against the square root of time in figure 2-5b. The effect of surface cover conditions on the rate of thaw penetration (hence the steepness of the plot) is most apparent in that line 1 was from Greenland (1939) under an insulating vegetation cover and line 4, showing most rapid thaw penetration, was from Spitsbergen (1974) and had no vegetation cover.

The active layer changes under the same climatic conditions according to many local parameters that influence the thermal conductivity of rocks. These are: geological structure, soil moisture, inclination and solar orientation of the slope, vegetation cover and snow. Carrying out my observations in Spitsbergen in 1978, I succeeded in examining experimentally the relation between the thickness of the active layer and one of these factors, slope inclination. Van Mijen Fjord (figure 2-4) turned out to be an ideal place for that purpose. There the slopes are extraordinarily uniform, the area being made up of dark Eocene slates. The mountains, reaching 1000 m in altitude, are cut by valleys and their slopes have varying inclinations — from a few degrees at the foot to over 35° on the gravitational slopes. A uniform slate weathering cover stretches over the whole surface of the slopes. The area is a true polar desert with no vegetation cover (figure 2-6).

The investigations were carried out in August 1978 by means of a metal sounding rod. The data (resulting from over 100 profiles) are displayed in figure 2-7. The greater the inclination of the surface, the smaller is the thawing of the soil. A thick active layer of about 70-100 cm was found at the foot of the mountains and on the flattenings of the slopes in the higher parts. There was no distinct dependence on altitude, as there was no marked

Fig. 2-6. Slopes of Van Mijen Fiord, uniform in structure but with different inclination, where measurements of the active layer thickness were undertaken, August 1978.

correlation between the solar orientation of the slopes and the thickness of the thawed layer, although the measurements were taken on slopes facing all directions.

The relation between the depth of summer soil thawing and the inclination of the surface was a dominant factor. Any possible influence that the height and solar orientation of the slopes might have had upon the development of the permafrost active layer was outweighed by the factor of surface inclination. The distribution of the points in the diagram indicates that there were deviations from the above-mentioned rule, but these can be explained by local factors.

In order to explain this specific phenomenon the heat conductivity of the slate weathering cover in summer has to be taken into consideration. This conductivity is the greater the more moist the weathering cover. The dry slate debris layer forms an excellent insulation against the downward penetration of the warming front. Wherever the weathering cover can retain

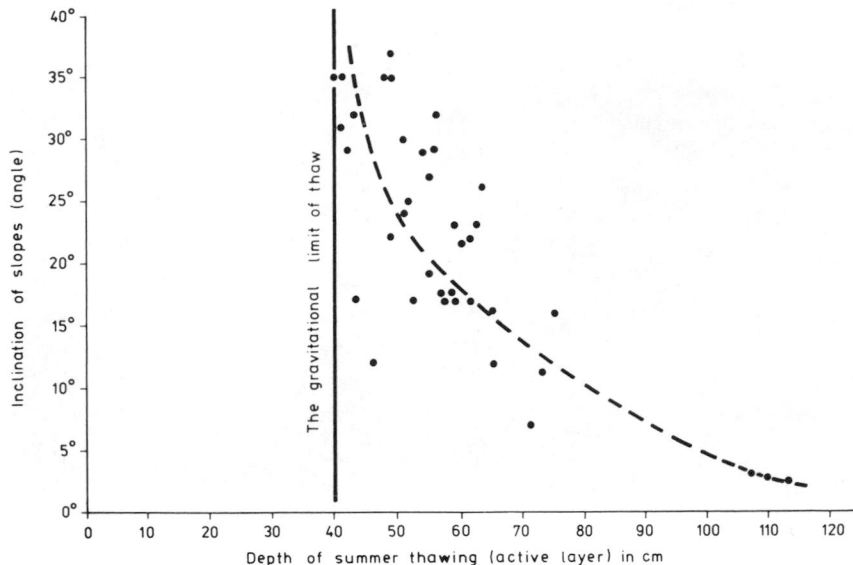

Fig. 2-7. Relation between the thickness of the active layer (summer thawing) and slope inclination in Van Mijen Fiord, in the summer of 1978. The smallest depth of thawing (40 cm) corresponds to the inclination of 35°. Some data points are means of several profiles.

water the conditions for heat penetration are favourable. This is mostly the case on slopes of small inclination, no matter whether surface depressions of this type are at the foot of mountains or in the middle of the slopes. Where the slopes are much inclined, the water flows down easily and the debris layer dries quickly, becoming a good insulating layer. Water comes mostly from thawing ice. Water derived from rainfall plays a very small role here because of its rare occurrence and rapid evaporation.

In cases where the weathering cover had a maximum inclination, i.e., one that corresponded to the angle of internal friction of the weathering material (gravitational slope), the active layer thickness in the investigated area was 40 cm. This thickness has been defined as the gravitational limit of thawing and means the maximum thickness of the debris layer capable of persisting on the slope in a state of relative equilibrium under the given climatic and weather conditions. This equilibrium is provided by the moisture content of the material which maintains a certain consistency of the whole layer. When the thickness of this layer increases, its surface becomes too dry for it to stay on the slope. The loose material slips down and the layer becomes reduced to its previous thickness.

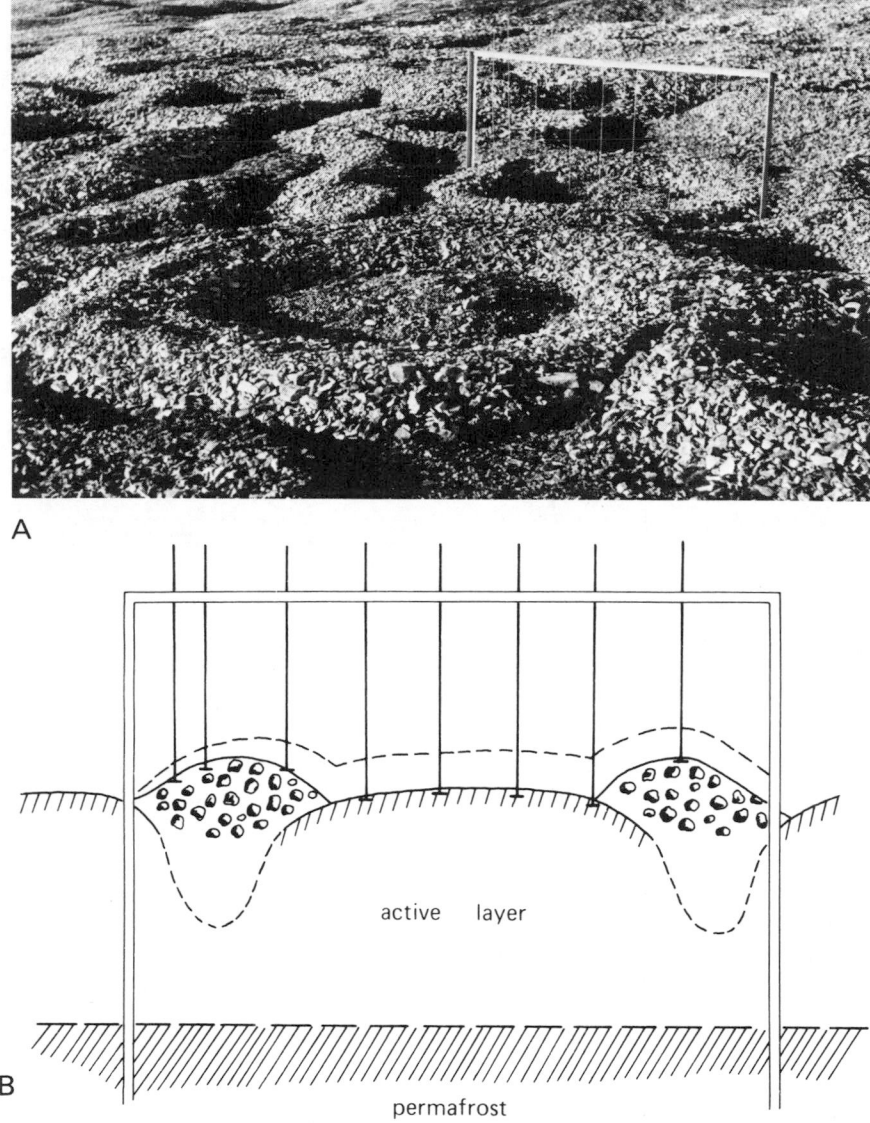

A

B

Fig. 2-8. (a) Sorted stone circles in Hornsund Fiord, Spitsbergen, one of them with a Bac motometer installed across its surface.

Fig. 2-8. (b) Bac motometer installed across a sorted stone circle. The broken line marks the position of the ground surface in winter.

EXPERIMENTAL RESEARCH ON PATTERNED GROUND AND GELIFLUCTION

It is known that frost factors cause high mobility of soil in polar areas. To observe the dynamics of these processes it was necessary to investigate soil movement both horizontally and vertically. I started my work in Greenland in 1937 (figure 2-3) but I did not obtain any results within one summer season. A turning point in this field came when Stanislaw Bac invented a very simple instrument known as a "Bac motometer" (Bac, 1952). In Poland the instrument was first used to measure frost soil movements at several agricultural research stations before World War II. It should be mentioned that instruments of this type were developed independently in other countries. The instrument described by Andrews (1963) was designed by the late Brian Haywood at the McGill Sub-Arctic Research Station at Schefferville, Quebec. The purpose of this "bedstead" was the same as that of the Bac motometer, namely to determine ground heave over a small area without disturbing the surface or snow cover.

In 1957, a motometer was installed in Spitsbergen in a field of sorted patterned ground (figure 2-8). For one year the vertical movement of soil was measured as illustrated in figure 2-9. The movement was greater in the middle of the field (amounting to 16 cm) but considerably smaller at the edges (5-6 cm). The lowest position of the ground surface was in early August, medium in September and highest in April. Winter frost heaving, the lowest position after ground thawing and, again, the late summer

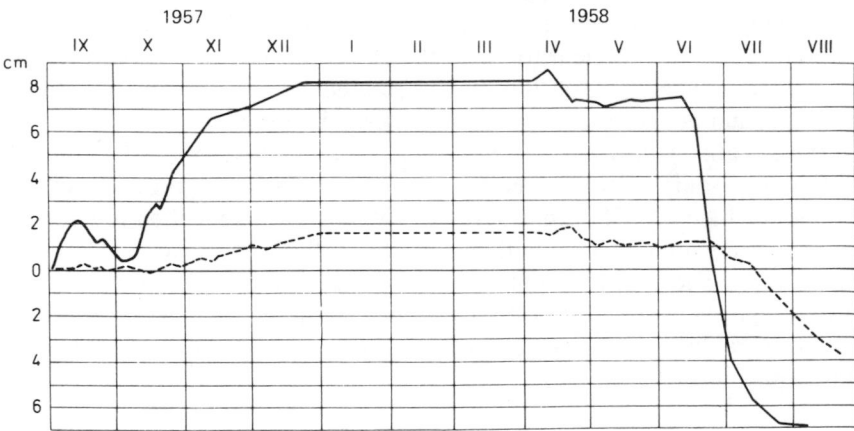

Fig. 2-9. Annual soil surface movement at the centre (continuous line) and edge (broken line) of a sorted circle in Hornsund Fiord, 1957-1958 (a graph of Czeppe (1961) and Jahn (1961), now modified and interpreted anew).

heaving of the soil surface could be distinguished clearly. In my report published immediately after the experiments (Jahn, 1961) I was not able to explain the late summer heaving of the ground. Today, after field experiments conducted by Parmuzina (1978) and Ross Mackay (1980) the phenomenon is known. Ground water close to the permafrost surface freezes in summer. The ice lenses that are formed push the soil upwards.

At the same time I measured the rate of frost heaving of soil, observing the position of wooden pegs inserted in soil to the depth of 60 cm. The result is shown in figure 2-10: the mean vertical movement of pegs in the first year

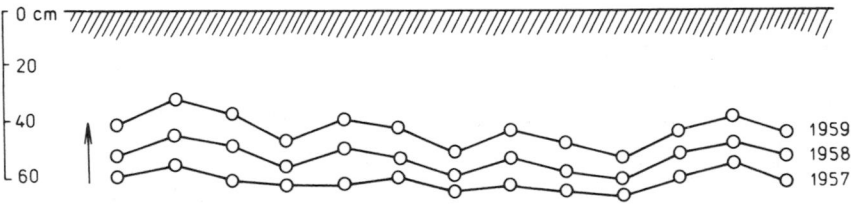

Fig. 2-10. Upfreezing of wooden pegs in Hornsund Fiord, Spitsbergen: the positions of the bases of pegs over the years 1957-1959.

was 7.8 cm, in the second 9.8 cm. The experiment proves that the process of upfreezing of the coarse and solid elements of soil (for example, stones) is one of the most dynamic periglacial processes. Thus, sorting processes and the development of sorted circles and polygons proceed very quickly. Two or three years are sufficient to make the effects of this process visible. In this way the experiments of 1957-59 contributed much to the explanation of the origin and development of sorted patterned ground.

At that time I also conducted observations to determine the rate of gelifluction (solifluction) movement. I inserted many peg profiles across gelifluction lobes. On the slope surface, with the average inclination of 10°, the maximum annual movement was 12 cm, with mean movement 3-4 cm (figure 2-11). These gelifluction measurements, which I started more than 25 years ago, were amongst the first experiments in this field. They gave the first measurable evaluation of gelifluction rate, which till then was assessed rather qualitatively.

THE ACTION OF WATER ON SLOPES

The question of slope wash by running water in a periglacial environment has always been open and opinions divided. Some authors have

A

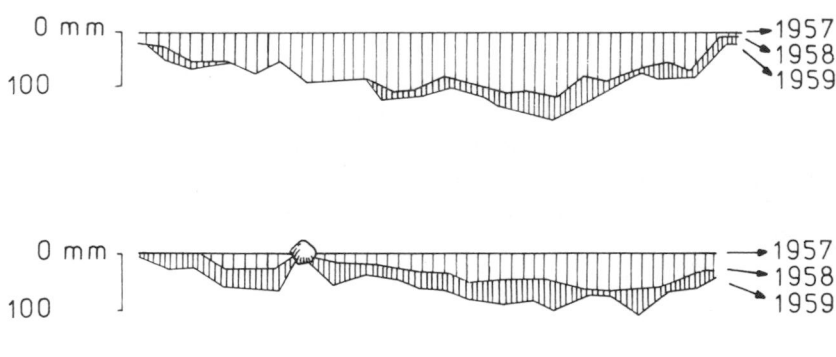

B

Fig. 2-11. Gelifluction measurement on the slopes of Hornsund Fiord, Spitsbergen.
(a) Deviation of pegs from the straight line, shown with a tape, took place within one year, in 1958. Black-coloured lower parts of pegs indicate annual thawing.
(b) Two years' displacement of the pegs installed across the gelifluction lobes, Hornsund Fiord, Spitsbergen.

recognized this geomorphological factor as the main force of erosion in this environment, whilst others even denied the existence of these processes.

To solve this problem, in 1957 and the years that followed I conducted field experiments near the Polish Research Station in Hornsund, Spitsbergen. Small plots of the slope surface were enclosed with boards and tin tanks 30 x 30 x 40 cm in size and pressed into the earth to collect runoff waters. Twice during each summer the material deposited by water in the tanks was removed (figure 2-12).

Fig. 2-12. Measurement of slope wash processes in Hornsund Fiord, Spitsbergen, July 1957.

The data collected at seven measuring points varied to a great extent. The denudation index calculated for one year was at maximum 16.3 g m^{-2}; on the average it was 5-8 g m^{-2} which gives 1 mm of slope denudation within 150-170 years. In my report I wrote:

> Such values may appear considerable. They indicate that under the present day climatic conditions prevailing in Spitsbergen running water is an important morphological factor. This is chiefly due to the fact that the slopes are unprotected by vegetation. A second cause is concentration of runoff during certain seasons. Melt waters, very abundant in spring, contribute chiefly to the destruction of the slope surfaces. Of lesser importance are the summer rain waters, though in Spitsbergen rains are sometimes intense during that season.

Which of these conclusions could be accepted today, after a 20-year period of further research? Observations of slope wash have recently focussed on Longyearbyen in Spitsbergen where, in July 1972, a rain storm of 31 mm rainfall in two days occurred. Many authors, like Thiedig and Kresling (1973), Thiedig and Lehmann (1973), and Rapp (1974), described effects of the storm, including numerous gullies on valley slopes. The denudation index calculated by them is 1 mm per year, that is, two orders of magnitude greater than the index I had calculated for the area of Hornsund. The problem, of course, centers on the question of the magnitude and frequency of operation of geomorphic processes. Rapp (1974) pointed out that the return period of the 1972 storm is greater than 53 years; how much greater is unknown. Contrary to the above mentioned authors I think, as I did 20 years ago, that the main factor of erosion is spring snow melt and not summer rain storms, because of the greater frequency of the former.

Considerable experimental field work remains to be done in the areas of permafrost, patterned ground, gelifluction and slope wash processes in periglacial environments. Ross Mackay demonstrates the way forward in his creative use of field experiments. As has been noted here, he is using and extending a method strongly rooted in European and especially Polish geographical, geological, pedological and engineering traditions.

REFERENCES

Andersson, J.G. 1906. Solifluction a component of subaerial denudation. J. Geol. *14*: 91-112.

Andrews, J.T. 1963. The analysis of frost-heave data collected by B.H.J. Haywood from Schefferville, Labrador-Ungava. Can. Geographer 7: 163-174.

Bac, S. 1952. Soil movements caused by action of frost (Polish with English summary). Panstw. Instyt. Geolog. Warszawa, Biul. *66*: 135-187.

Brown, J., Haugen, R.K. and Parrish, S. 1975. Selected climatic and soil thermal characteristics of the Prudhoe Bay Region. *In* J. Brown, ed., Ecological Investigations of the Tundra Biome in the Prudhoe Bay Region, Alaska, Biol. Papers, Univ. Alaska, Spec. Rep. *2*: 215 pp.

Brown, J., Rickard, R. and Victor, D. 1969. The effects of disturbance on permafrost terrain. U.S. Army Cold Regions Research and Eng. Lab., Spec. Rep. *138*: 13 pp.

Czeppe, Z. 1961. Annual course of frost ground movements at Hornsund, Spitsbergen, 1957-1958 (Polish with English summary). Zesz. Nauk. Uniwer. Jagiellonskiego, Krakow, z.3, Prace Inst. Geogr., z.25.

Högbom, B. 1914. Uber die geologische Bedeutung des Frostes. Uppsalas Univ., Geol. Inst. Bull. *12*: 257-389.

Jaczewski, .A. 1889. O vejacnomerzloj pochve v Sibiri i ledyanykh slojakh (On permafrost soil in Siberia and ice layers). Izw. Rusk. Geogr. Obshchestva *25*: v. 5.

Jahn, A. 1946. Badania nad struktura i temperatura gleb w Zachodniej Grenlandii (Researches on structure and temperature of the soil in Western Greenland). Rozprawy Wydz. Mat. -Przyr. Akademii Umiejetnosci, Krakow, *72*: 1-122.

Jahn, A. 1954. Services of Walery Lozinski in the field of periglacial studies. Biul. Peryglacjalny *1*: 7-18.

Jahn, A. 1961. Quantitative analysis of some periglacial processes in Spitsbergen. Zesz. Nauk. Uniwer. Wroclawskiego, Ser. B, *5*: 54 pp.

Jahn, A. 1975. Problems of the Periglacial Zone. Polish Scientific Publishers, Warszawa: 219 pp.

Leffingwell, E.K. 1919. The Canning River Region, Northern Alaska. U.S. Geol. Surv., Prof. Paper *109*: 251 pp.

Loziński, W. 1909. Über die mechanische Verwitterung der Sandsteine im gemässigten Klima. Acad. Sci. Cracovie, Bull. Internat., Cl. Sci. Math. et Nat. *1*: 1-25.

Mackay, J. Ross. 1980. The origin of hummocks, western Arctic coast, Canada. Can. J. Earth Sciences, *17*: 996-1006.

Meinardus, W. 1912. Beobachtungen über Detritussortierung und Strukturboden auf Spitzbergen. Gesell. Erdkunde Berlin, Zeitschr. (1912): 250-259.

Parmuzina, O. Yu. 1978. The cryogenic structure and certain features of ice separation in a seasonally thawed layer (in Russian). *In* A.J. Popov, ed., Problemy kriolitologii (Problems of cryolithology). Moskva, Izdatel'stvo Moskovskogo Universiteta 7: 141-163.

Rapp, A. 1974. Slope erosion due to extreme rainfall, with examples from tropical and arctic mountains. Geomorphologische Prozesse und Prozesskombinationen in der Gegenwart unter Verschiedenen Klimabedingungen. Nachr. Akad. Wiss. Göttingen, III. Math.-Phys. Kl., Nr. *29*: 118-336.

Ryden, B.E. and Kostov, L. 1980. Thawing and freezing in tundra soils. *In* M. Sonesson, ed., Ecology of a Subarctic Mire. Ecol. Bull. Stockholm, *30*: 251-281.

Sumgin, M.J. 1937. Vechnaya merzlota pochvy v predelakh SSSR (Permafrost of the soil within the USSR boundaries). Izv. Akad. Nauk. SSSR, Moskva. 2nd. ed.: 379 pp.

Terzaghi, K. 1952. Permafrost, Boston Soc. Civil Eng., J. *39*: 319-368.

Thiedig, F. and Kresling, A. 1973. Meteorologische und geologische Bedeutungen bei der Entstehung von Muren im Juli 1972 auf Spitzbergen. Polarforschung *43*: 40-49.

Thiedig, F. and Lehmann, V. 1973. Die Entstehung von Muren als säkulares Ereignis auf Spitzbergen (Svalbard) und ihre Bedeutung fur die Denudation in der Frostschuttzone. Mitt. Geol.-Palaont. Inst. Univ. Hamburg *42*: 71-80.

Troll, C. 1944. Strukturboden, Solifluktion und Frostklimate der Erde. Geol. Rundschau *34*: 545-694.

Walker, H.J. and Harris, M.K. 1976. Perched ponds: an arctic variety. Arctic *29*: 223-238.
Washburn, A.L. 1956. Classification of patterned ground and review of suggested origins. Geol. Soc. Amer. Bull. *67*: 823-865.

3
EXTREME RAINFALL AND RAPID SNOWMELT AS CAUSES OF MASS MOVEMENTS IN HIGH LATITUDE MOUNTAINS

Anders Rapp

Department of Physical Geography, The University of Lund
S-223, 62 Lund, Sweden

ABSTRACT

Debris slides, debris flows and slush avalanches are significant episodic mechanisms of denudation in periglacial mountain regions. In this paper, an attempt is made to assess their magnitude and frequency in the long term in northern Scandinavia. Debris slides and flows are initiated by intense summer rainfall. Rainfall impact is classified according to the area affected: 'regions', 'cells' and 'spots' are defined on a progressively narrowing local scale. Relevant events in this century are mapped and tabulated. Lichen measurements are investigated as a means to extend the record to over 1,000 years. It is concluded that debris slide and debris flow activity has been continuous through Holocene times with a return period of 50-200a for major events yielding specific denudation in the order of 1 to 10B. Slush avalanches are relatively frequently repeated (return periods between several years and some decades) but are restricted to fixed tracks. Because of their frequency, their geomorphic importance is comparable with that of debris slides and flows. The two classes of events occur at different times of year and present distinct hazards: because of this they should each be studied more thoroughly.

RÉSUMÉ

Pluie abondante et fonte de neige rapide comme causes des mouvements de masse dans les montagnes de haute latitude

Les glissements de débris, les coulées de débris, et les avalanches de neige sursaturée constituent les mécanismes épisodiques de dénudation dans les régions périglaciaires de montagnes. Dans cet article, on tente d'évaluer leur ampleur et leur fréquence à long terme dans le nord de la Scandinavie. Les glissements et coulées de débris sont déclenchées par des pluies intenses en été. L'efficacité de la pluie est classée selon la superficie affectée: 'région', 'localité' et 'endroit' sont définies selon une échelle progressivent plus petite. Les évenements du siècle présent sont relevés et cartographiés. Des mensurations lichenométriques sont utilisées pour étendre les observations jusqu'à plus de mille années. Les glissements et les coulées de débris ont existé durant tout l'Holocène avec une périodicité de 50 à 200 ans pour les évenements les plus importants, ce qui aboutit à une dénudation spécifique de l'ordre de 1 à 10 B. Les avalanches de neige sursaturée se répètent assez fréquemment (périodicité de plusiers années à quelques décennies) mais elles sont limitées à des tracés fixes. Vue leur fréquence; leur importance géomorphologique est comparable à celle des glissements et coulées de débris. Les deux catégories d'évènements se produisent à des temps différents de l'année et présentent des hasards distincts: de ce fait, elles devraient être étudiées en plus grand détail.

ПРОЛИВНЫЕ ДОЖДИ И БЫСТРОЕ ТАЯНИЕ СНЕГОВ КАК ПРИЧИНА ДВИЖЕНИЯ ГРУНТА В ГОРАХ НА ВЫСОКИХ ШИРОТАХ

А. РАПП

РЕЗЮМЕ

Обломочные оползни и потоки, и лавины талого снега, являются важными эпизодическими механизмами денудации в перигляциальных районах гор. В настоящем докладе делается попытка оценить их размер и частоту в долговременном масштабе в северной Скандинавии. Оползни и потоки обломочного материала побуждаются проливными летними дождями. Влияние дождя подразделяется согласно территории подвергающейся его действию: »районы«, »клетки«, »точки« прогрессивно локализируются. События настоящего столетия представлены на карте и в форме таблиц. Изучаются размеры лишайника, чтоб таким образом расширить применимость данных до тысячи лет и дольше. Выводится заключение, что оползни и потоки продолжались в течение всего Голоцена, с периодом повторяемости важных событий от 50 до 200 лет, приводящие к денудации в размерах от 1 до 10 Б. Лавины талого снега повторяются довольно часто (период повторяемости от нескольких лет до нескольких десятков лет), но они ограничиваются до установленных трасс. В виду их частой повторяемости, их геоморфическое значение равнозначно обломочным оползням и потокам. Эти два подразделения событий случаются в разное время года и представляют собой определенную опасность; поэтому каждое из них следует изучать с большим вниманием.

INTRODUCTION

In the 1950s the author carried out field studies of contemporary geo-morphological processes in the high mountains of northern Scandinavia, particularly in the trough valley of Kärkevagge west of Abisko in northern Sweden (Rapp, 1960). The monitoring of slope processes active in the period 1952-1960 was summarized by calculating the mass transfer of material downslope, expressed in tonnes times metres of vertical component of movement. The ranked list of mass transfer processes is given in Table 3-1.

TABLE 3-1
RANKED LIST OF MAJOR SLOPE PROCESSES
IN KÄRKEVAGGE, LAPPLAND, 1952-1960
(*Modified from Rapp, 1960: 185*)

Process	Tonnes km^{-2} a^{-1}	Denudation mm a^{-1}	Mass transfers Tonne-metres	Remarks
1. Transport of solutes by running water	26	0.010	136,500	
2. Debris slides and flows	49.4	0.019	96,300	extreme event, October, 1959
3. Slush avalanches, rock debris transport	14	0.005	20,000	extreme events, 1956 and 1958
4. Rockfalls	8.7	0.003	19,600	seasonal events of high frequency
5. Solifluction	5.4	0.002	5,300[a] 19,800[b]	Based on 9 km of soliflucted slope length; material density of 1.8
6. Talus creep	1.5	0.001	2,700[a] 4,700[b]	Based on 6 km of talus slope length; material density of 1.8

(a) *Vertical component.*
(b) *Horizontal component.*

Transportation by solutes in running water was the most important geomorphic process in terms of mass transfer but debris flows and slides were most important in terms of specific denudation over the 9 year period. How important is the latter group of processes during longer time periods and in other lithologies? The debris slides and flows of October 1959 in Kärkevagge were regarded by the author as ''a centennial or probably even millennial maximum'' (Rapp, 1960, p. 185). This paper will discuss the geomorphic importance of debris flows, slides and slush avalanches in high latitude mountains. The concepts of ''magnitude and frequency'' of geomorphic processes (Wolman and Miller, 1960), ''extreme events'' (e.g.,

Starkel, 1976), and "thresholds in geomorphology" (Coates and Vitek, 1980) are helpful conceptual backgrounds to the discussion.

DEFINITIONS OF DEBRIS SLIDES, DEBRIS FLOWS, AND SLUSH AVALANCHES

General classifications of mass movements are based on material properties and the type of movement (cf. Sharpe, 1938; Varnes, 1958; Zaruba and Mencl, 1968). To these two diagnostic criteria can be added the morphology and fabric of the deposits, which can help considerably in the reconstruction of mass movement.

Three major classes of rapid slope processes are distinguished: fall, slide and flow. Fall is the free falling through the air down a cliff, or the rolling or bouncing of individual blocks of rock or other debris down a steep slope. Debris slide is a rapid movement of a mass of rock or debris, gliding on one or several slide planes on the substratum and causing considerable frictional erosion. Debris slides are characterized by a distinct slide scar and an eroded slide track terminating in a slide tongue or lobe of overthrust, folded and deformed slabs. However, many debris slides, which are heavily oversaturated by water, turn into a movement of debris flow, viscous or plastic, and create lateral debris flow levees which end down-slope in a frontal lobe or in an outspread debris flow fan (figures 3-1 and 3-2).

The term "avalanche" I restrict to mass movements of snow. They can be mixed with rock and soil debris and, if so, they have been called "dirty avalanches" (Rapp, 1960, p. 127) or "mixed avalanches" (Washburn, 1979, p. 193). Slush avalanches are a particular type of snow avalanche that occur along steep watercourses of mountain brooks. They are released by meltwater that saturates the thick snow cover in small streams. They consist of large masses of very wet and heavy snow, ice blocks, water, and often include large quantities of eroded soil and rock. They mark the catastrophic spring break-up event in small mountain streams.

MAGNITUDE AND FREQUENCY OF ALPINE DEBRIS FLOWS IN SCANDINAVIA

Figure 3-3 is a map summing up the spatial pattern of known rainfall triggered mass movements in the mountain area of Abisko and Kebnekaise during the period 1952-1980. Three areal size classes of rainfall impact appear.

(1) An area of more than 6000 km covered by the frontal cyclonic rainfall of October 5-6, 1959 was marked by mass movements on steep slopes from the Norwegian coast to the vicinity of Abisko. This we call a rainfall impact *region*.

(2) Two areas occur of 11 and 50 km^2 size: Tarfala (1972: figure 3-4), and

Fig. 3-1. Fresh light-coloured debris flow levees of 1979 deposited on talus slopes of dark amphibolite debris, Nissunvagge, Section C. Note also deposits of high blocky lobes at talus base. In foreground tundra with sub-recent ice-wedge polygons. (A.R. 1980.)

Nissunvagge (1979). Such areas we call rainfall impact *cells*. They are caused by local rainfall of high intensity, as in convective cells.

(3) There are three spots with one or a few mass movements triggered by intensive rainfall. These are at Kaisepakte (1956), in Kärkevagge (1975) and at Torneträsk station (1979). We call these rainfall impact *spots*.

The inventory is not complete. Only the areas most thoroughly observed, like those near the railway line or near the research stations at Abisko and Tarfala, are reasonably well covered by observations. Many single mass-movement spots and some cells may have escaped observation. However, the pattern points to the long term importance of rainfall-triggered mass movements and flooding effects on the geomorphology of these northern mountains.

A number of cases of debris flows and related phenomena triggered by extreme rainfall in the mountains of Scandinavia and Spitsbergen have been listed in Table 3-2. They have been numbered from 1 to 10, with 1a and 1b occurring on the same day, and likewise 7a and 7b. The apparently higher

TABLE 3-2

REPORTS OF ALPINE DEBRIS SLIDES AND FLOWS
TRIGGERED BY HEAVY RAINS IN SCANDINAVIA AND SPITSBERGEN

Nr	Location	Date	Rainfall and duration	Area[a] affected	Number of events	Reference
1a	Nissunvagge, N. Sweden	23.6.79	Unknown, 1 hr	C	>200	Rapp and Nyberg, 1981
1b	Torneträsk stn, N. Sweden	23.6.79	18 mm	S	1	Rapp and Nyberg, 1981
2	Lyngseidet, N. Norway	6-7.8.75	Unknown, 17 hrs	C	10	R.G. Healey, Edinburgh (pers comm)
3	Kärkevagge, N. Sweden	10.8.75	Unknown	S	1	Rapp and Strömquist, 1979
4	Tarfala N. Sweden	7.7.72	45 mm, 2 hrs	C	20	Rapp, 1974
5	Longyearbyen, Spitsbergen	10-11.7.72	31 mm, 12 hrs	R	50	Rapp, 1974
6	Ulvådal W. Norway	26.6.60	Unknown, 1 hr	C	30	Rapp, 1963
7a	Kärkevagge, N. Sweden	6.10.59	107 mm, 12 hrs	R	33	Rapp, 1960
7b	Andöya, N. Norway	6.10.59	52 mm, 12 hrs	R	20	Strömquist, 1976
8	Kaisepakte, N. Sweden	8.7.56	(23 mm)	S	1	Rapp, 1960
9	Ajaure, Tarna, N. Sweden	24.6.47	Unknown	C	10	Rudberg, 1950
10	Arjeplog N. Sweden	3.8.20	Unknown	S	1	Zenzén, 1926

a   Area affected is indicated by letters R = region, C = cell, S = Spot.

frequency of events in the 1970s probably reflects closer observation and reporting of events than earlier.

The dates of the events are from June 23 to October 6, with the following distribution by month: June (3 cases), July (3), August (3), September (0), October (1). The period of main activity coincides with the summertime precipitation maximum in interior northern Scandinavia (see table 3-3). Rainfall magnitudes are difficult to analyse, since the time period of rainfall recording is generally 24 hours and the rainfall measurements are not directly from the area affected, except in the cases of Tarfala and Long-yearbyen in 1972. However, rainfall from 18 mm (case 1b) to 107 mm (case 7a) was recorded within 5 km of the debris flows.

Several of the descriptions of mass movements triggered by rainstorms in the Scandinavian mountains also contain references to old events of the

Fig. 3-2. Detail of debris flow levees of 1979 (foreground) over gently sloping meadow at the base of cone C42, Nissunvagge. Fresh debris flow lobes in background. (A.R. 1980)

same type, either from historical tradition or from local vegetation or geomorphological evidence. Thus, a debris slide near Arjeplog in Swedish Lappland in 1920 was preceded by a similar event 70 years earlier (Zenzén, 1926, cf.figure 3-5). Rudberg (1950, p. 148) reported cases of mass movements in Kittelfjäll, where the preceding events occurred 35 years earlier. The debris slide in 1959 in an area at Andöya, North Norway was affected

TABLE 3-3

MONTHLY MEAN VALUES OF TEMPERATURE AND PRECIPITATION
AT RIKSGRÄNSEN AND ABISKO,
REDUCED TO THE PERIOD 1901-1930 (FROM WALLÉN, 1960)

Station	Jan	Feb	Mar	Apr	May	Jun	Jul	Aug	Sept	Oct	Nov	Dec	Year
Temperature (°C)													
Riksgränsen	-9.7	-10.5	-8.6	-4.1	+0.7	+6.2	+10.6	+9.4	+4.2	-1.6	-5.7	-9.2	-1.5°
Abisko	-10.0	-10.8	-8.0	-3.0	+1.9	+7.4	+11.1	+9.7	+5.2	-0.7	-5.4	-9.2	-1.0°
Precipitation (mm)													
Riksgränsen	90	61	63	60	51	70	69	66	109	64	76	64	844
Abisko	20	15	14	11	16	30	42	38	28	17	17	19	267

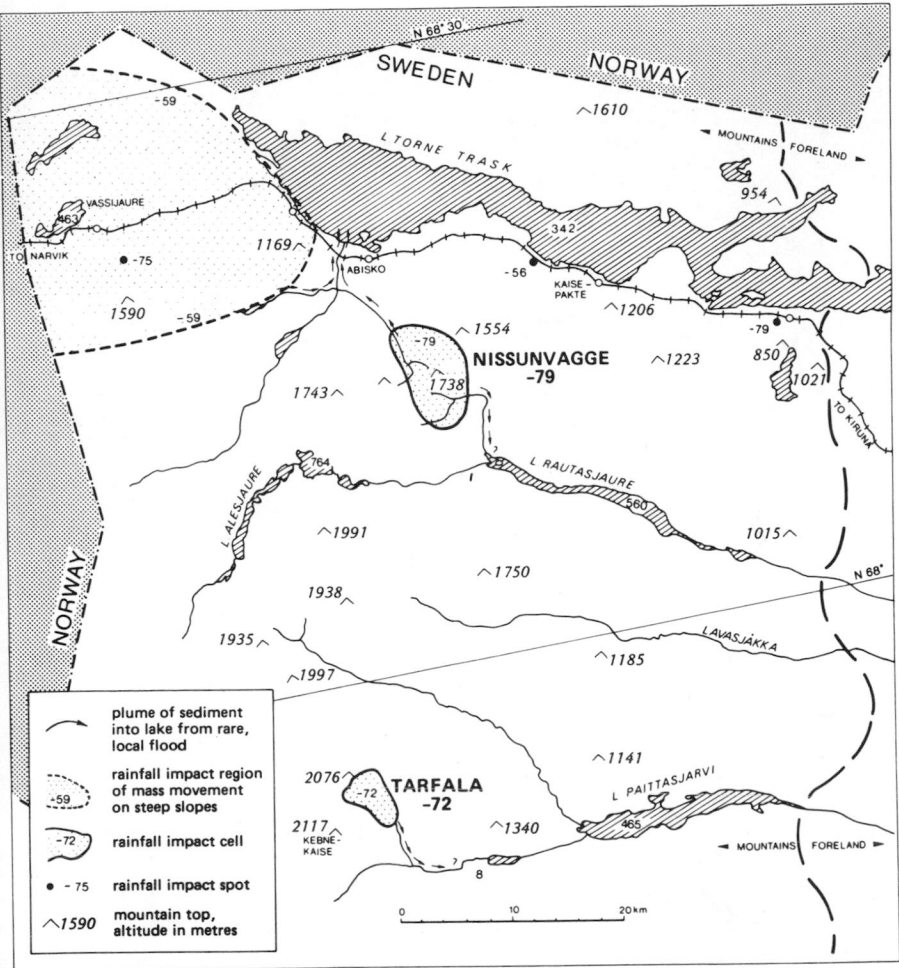

Fig. 3-3. Overview map of reported cases of alpine debris flows and slides triggered by intensive rainstorms in the mountains of Abisko and Kebnekaise in the period of observation 1952-1980.

by similar erosion 61 years earlier and probably also 110 years earlier (Strömquist, 1976).

Analyses of maximum 24-hour rainfall do not contradict the evidence of historical sources, vegetation dating or geomorphological evidence of repeated occurrences of debris flows and slides of 50-200 years return periods in the high mountains of northern Scandinavia. We think that in

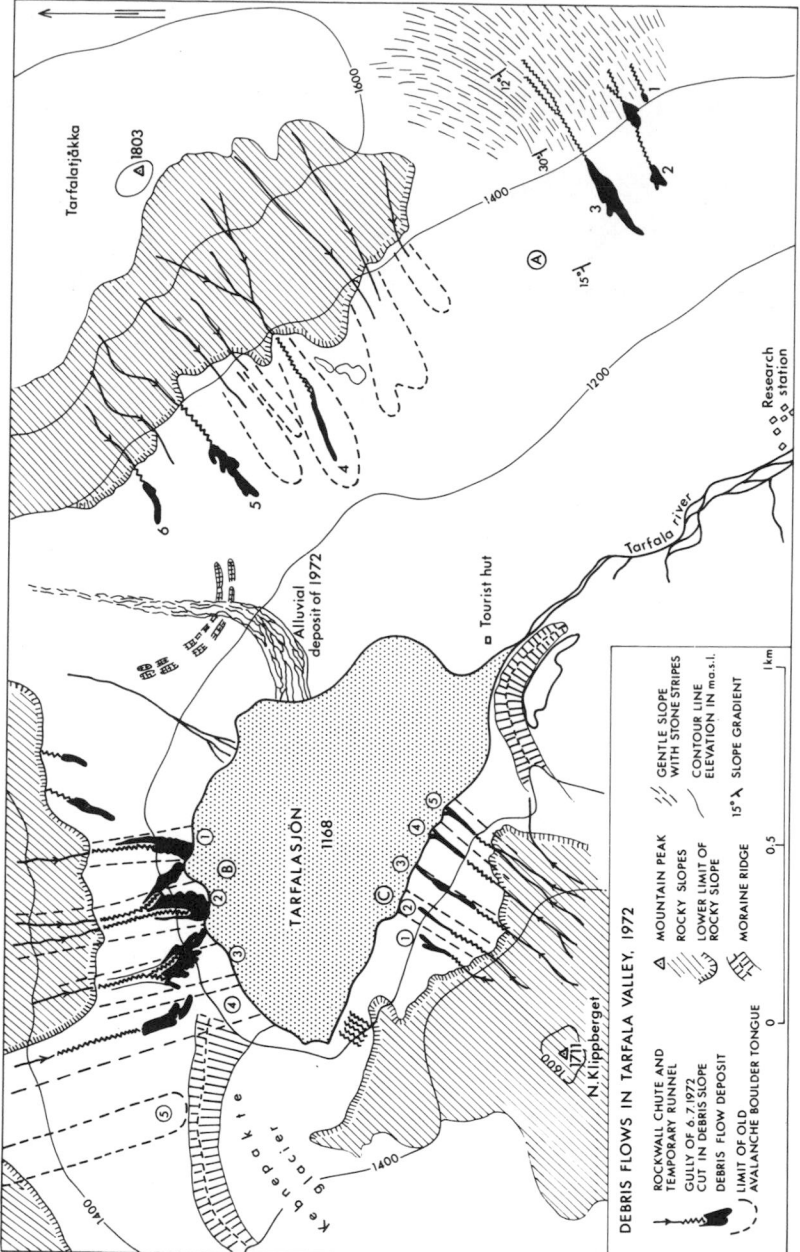

Fig. 3-4. Map of debris flows of July 6, 1972 in Tarfala, illustrating a rainfall impact cell. 45 mm of rainfall were recorded at the Research station during two afternoon showers. Debris flow deposits varied in size from small levees of some tens of m^3 (e.g. C1 - C5) to large diverging, blocky, flow lobes of several thousands of m^3 (B1 - B4).

these and other periglacial alpine conditions the dating by lichens is a particularly promising method for further investigation.

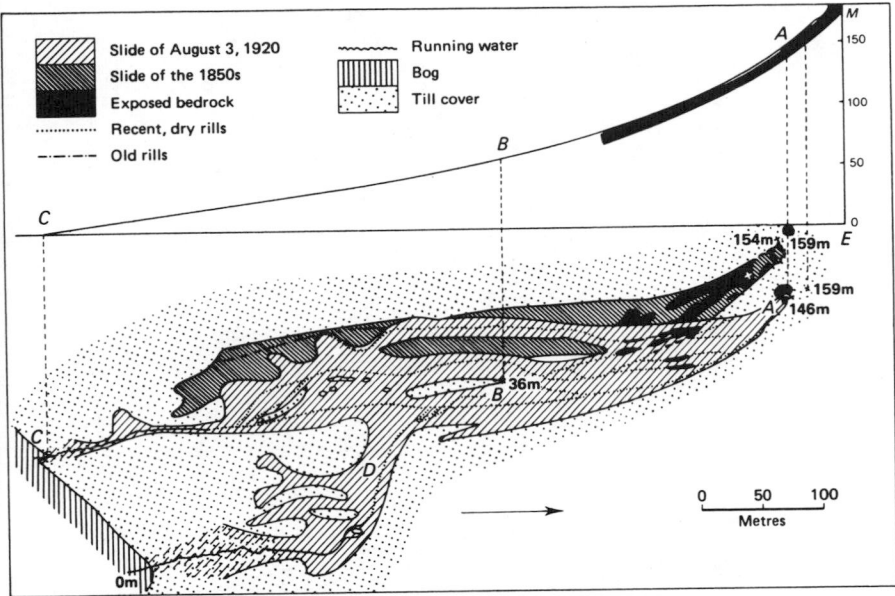

Fig. 3-5. Map of two generations of debris slides triggered by convective rainstorms in Arjeplog, N. Sweden. (From Zenzén, 1926.)

DATING BY LICHEN MEASUREMENTS

The debris flow cone (colluvial cone) C 42 in Nissunvagge (figure 3-6) was selected for a study of the old debris flow features and the possibilities of dating by the use of lichenometry. The photograph and the drawing of C 42 illustrate the situation and the micromorphology. The two diverging fresh debris lobes of C 42 are deposited on the lower part of a colluvial cone and on alpine meadows beyond the base of that cone. The right half of the cone is unaffected by recent debris flows but its right margin is marked by a pair of old debris flow levees. A whole complex of old, bouldery front lobes is visible between and beyond the distal parts of the two lobes of 1979 (see figure 3-6).

The measurements of diameters of lichens of the two species *Rhizocarpon geographicum* and *Rhizocarpon alpicola* followed the method described by Karlén (1973) for the dating of moraines in the Kebnekaise mountains. Inventories of surviving *Rhizocarpon* specimens on fresh lobes and levees of the 1979 deposits were undertaken. Almost all boulders were

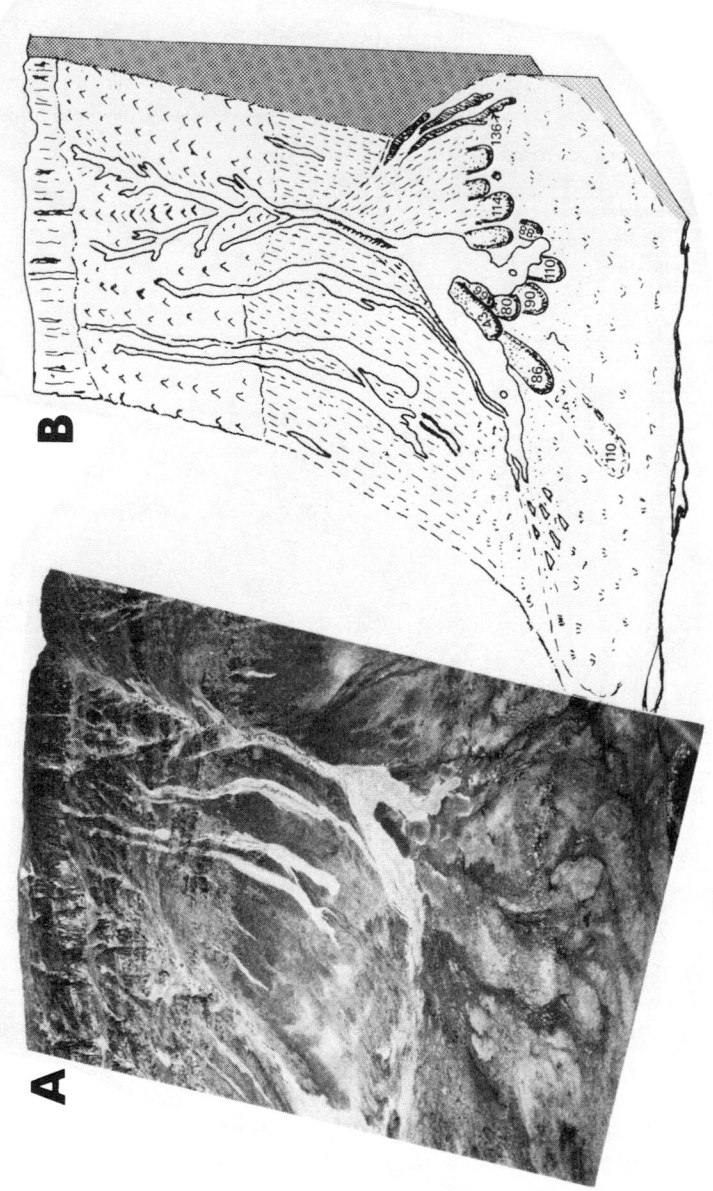

Fig. 3-6. Photograph (A) and block diagram (B) of Mt. Nissuntjarro with fresh debris flows C38 to C42, of 1979. Old debris flow lobes and levees of colluvial cone C42 (right) with figures showing maximum diameter of *Rhizocarpon* lichens, in mm. Summit plateau at 1600 m, base of new debris flow at 1000m. (Photo: N.Å. Andersson, July 25, 1979. Drawing: Rapp and Nyberg, 1981.)

TABLE 3-4

MAXIMUM LICHEN DIAMETERS ON DEBRIS FLOW LOBES
OF CONE C42 IN NISSUNVAGGE, AND ESTIMATES OF THEIR AGE
(from Rapp and Nyberg, 1981)

Maximum lichen diameter, mm	Position at C42	Corresponding age
136	large levees	1 500 a
114	large lobe at right of centre	900
110	lobe near centre	700
90		380
80	4 overlapping generations near centre	280
66		220
43		120
110		700
86	3 overlapping generations to the left	330
43		120

free from lichens and clean of any living vegetation due to the scouring and grinding of boulders during the mass movements. We made linear inventories of 300 boulders each on several deposits and discovered 0 to 4% of boulders with surviving and exposed *Rhizocarpon*. The higher percentage was on the outer side of levees. In order to eliminate the possible error of including a boulder with surviving lichens in our measurements of maximum lichen diameter on old deposits, the maximum specimen of 20 counted and measured was discarded if it was at least 10 mm larger than the second largest lichen. The biggest lichen on twenty separate boulders was measured. The ages as given in Table 3-4 are from Karlén's (1973) curve for Kebnekaise-Sarek. We stress that we are presenting only a preliminary attempt to date different episodes of debris flow by lichen measurements, which should be completed by larger numbers of measured lichens and also extended to other colluvial cones in Nissunvagge.

The following tentative conclusion can be derived from the evidence of figure 3-6:

(1) The 1979 debris flows in Nissunvagge are not unique. There are many old debris flows appearing on the deposits of colluvial cones and also on simple debris slopes. The predominant pattern of cone C 42 is the cumulative build-up of the cone's distal part by repeated debris flows, of which at least five overlapping events during the last 700 years are possible, and at least two older events to 1500 years BP are likely.

(2) The bulk of the colluvial cones and debris sheets has been formed by repeated Holocene events, not by flows at the deglaciation 8000 to 9000 years ago.
(3) The large size and downslope extent of the debris flows make it likely that the old debris flows have been triggered by rainstorms in summer on unfrozen ground — not by snowmelting.
(4) The influence of deposits of running water, rockfalls and slush avalanching is of smaller importance than that of debris flows on many of these slopes.

For comparison we also refer to the study of a similar case made by R. Curry (1966) in the Tenmile Range, central Colorado, USA. Curry observed a series of alpine debris flows after a rainfall of 245 mm in 24 hours. The flows occurred as a series of pulses over and through saturated talus on slopes as steep as 41°. Four successive ancient deposits of debris flows were recognized by Curry and his lichenometric analyses indicated that they occurred about once every 150-400 years.

ESTIMATION OF RATE OF DENUDATION BY DEBRIS FLOWS

The rate of rock denudation can be given in mm per 1000 years, or so called "Bubnoff units" (B) (Fischer, 1969). In table 3-5 data of "debris

Fig. 3-7. The runout zone of a slush avalanche on the ice of Lake Rissajaure, Kärkevagge, on June 12, 1956. The avalanche deposit contains 200 m^3 of rock debris, mixed with snow. (A.R. 1956.)

denudation" and of "rock denudation" have been listed. Rock denudation data are here supposed to be 70% of debris denudation, due to a reduction of an assumed porosity of 30% in the debris masses. In the table we have made a number of extrapolations and assumed that the recurrence interval of major debris flow events is 200 years, as indicated by the evidence from Nissunvagge (table 3-4). The extrapolations result in a very wide range of denudation rates, from 1B in the case of Kärkevagge (1959) to 150 B in the case of the debris slides of Ulvådal in 1960. It is evident that the internal factors, like rainfall intensity (cf. Coates and Vitek, 1980), are probably very different.

TABLE 3-5

CASES OF ALPINE DEBRIS SLIDES AND FLOWS
IN SCANDINAVIA AND SPITSBERGEN
AND ESTIMATED INDEXES OF THEIR DENUDATION IMPACT.

Locality	Date	Catchment area km^2	Rock type	Volume of debris m^3	Denudation Debris mm	Rock mm	Extrapolated rock[a] mm
Nissunvagge N. Sweden	23.6.79	5.5	amphibolite	85 000	15.5	10.8	54
Tarfala N. Sweden	6.7.72	11	amphibolite	55 000	5	3.5	18
Kärkevagge N. Sweden	6.10.59	15	mica-schists	4 600	0.3	0.2	1
Longyearbyen Spitsbergen	11.7.72	4.5	schists, sandstones	5 000	1	0.7	3.5
Ulvådal W. Norway	26.6.60	7	granite	300 000	43	30.1	150

a. Extrapolated rates of rock denudation are given in mm per 1,000 years assuming that the recurrence interval is 200 years.

Morphological evidence from Ulvådal in the form of erosion and removal of a 10,000 year old till deposit in 1960, and from a 5.5 km area of Nissunvagge in the form of exceptionally dense debris flow development in 1979, indicates that the assumption of a 200 year recurrence interval is unrealistic for some monitored events. It seems probable that the long-term contribution of debris slides and flows to denudation is less than 10B.

SLUSH AVALANCHES

Erosive slush avalanches have been described as a distinct slope process in the mountains of Scandinavia and Spitsbergen. In this report figures 3-7 to 3-10 show typical features of slush avalanches and their sites. They occur at the interval of some years and follow steep, first order streams from hollows or chutes favourable for the accumulation and storing of snow, such as cirques (figure 3-10). Slush avalanches consist of masses of wet, heavy snow, ice, rock debris, soil and vegetation debris.

Characteristic surface features of the tracks and deposits of slush avalanches are striations and grooves in the vegetation and soil from removed and dragged rocks (e.g. figure 3-9, zone "e"). Other typical detail forms are avalanche debris tails (Rapp, 1959), which are wedge-like ridges of rock fragments on the downslope side of fixed larger boulders. The boulder tongues of slush avalanches are generally longer and flatter than those deposited by other snow avalanches. They have a frontal, horseshoe-shaped fringe of perched boulders on low slope gradients, down to zero inclination or even crossing the stream in the valley bottom and reaching somewhat up the opposite slope. Corner (1980) has recently drawn attention to such features under the particular names of avalanche impact tongues, pits and pools, in a study based on interpretations of aerial photographs and field work in the Tromsö region, North Norway.

An ongoing study in the Abisko mountains by Nyberg (1980) shows a higher frequency of slush avalanches than has hitherto been assumed. Recurrence intervals seem to vary from only a few years up to several decades at different sites (cf. figure 3-11). The railway line from Kiruna, Sweden, to Narvik, Norway, is affected by avalanches particularly at three sites: Mt. Nuolja, west of Abisko; Mt. Kaisepakte (figure 3-11, site no. 16); and Kvitur in Norway.

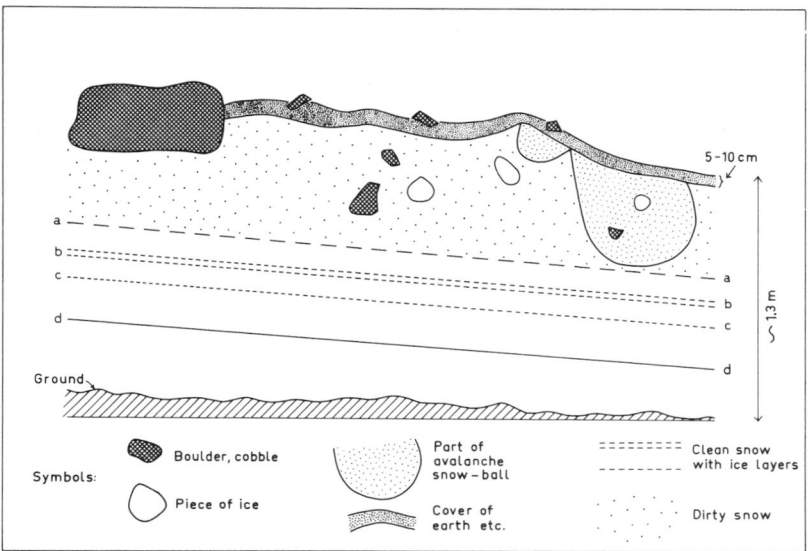

Fig. 3-8. Vertical section through remnants of slush avalanche after 5 weeks of melting, Kärkevagge, V45, 1958.

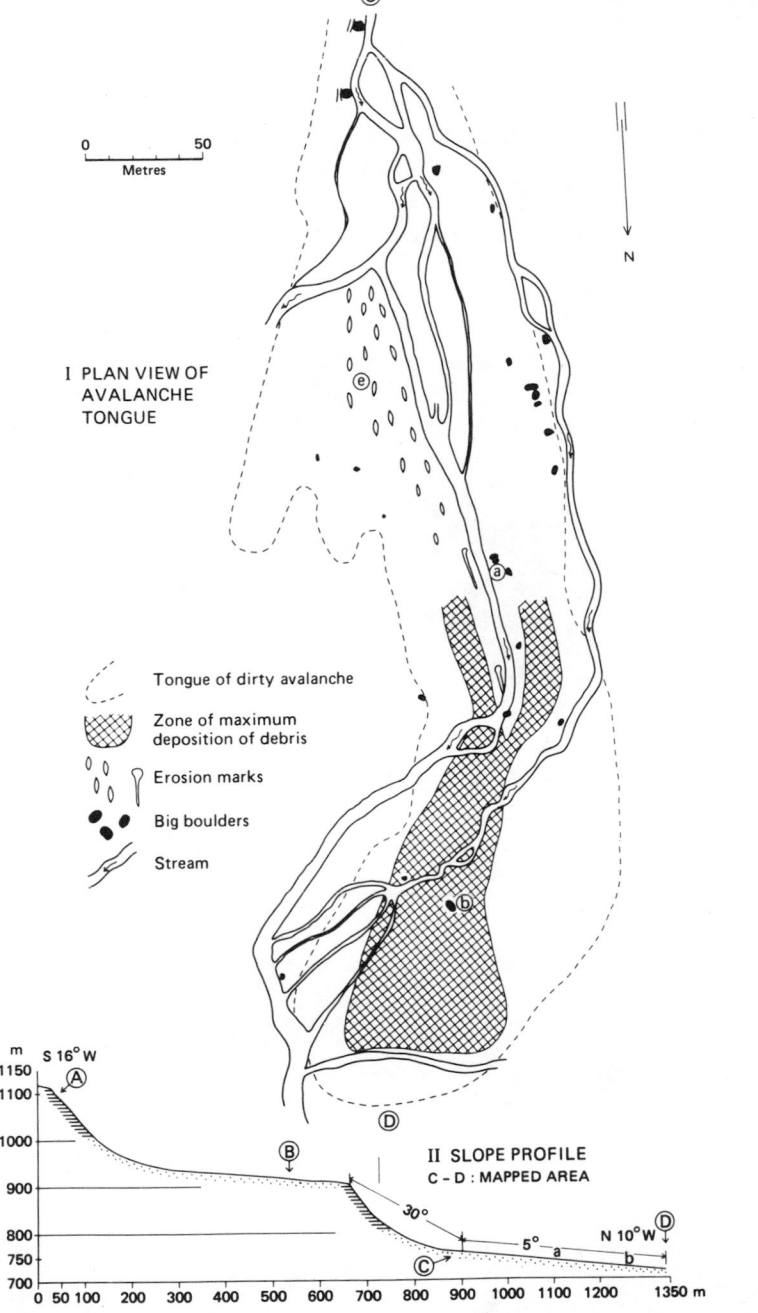

0   50
Metres

N

I PLAN VIEW OF
AVALANCHE
TONGUE

e

a

- - - - Tongue of dirty avalanche

Zone of maximum
deposition of debris

Erosion marks

Big boulders

Stream

b

m   S 16° W
1150
1100
A
1000
B   II SLOPE PROFILE
          C – D : MAPPED AREA
900
D
30°
800   5°   a   N 10° W   D
750
C
700
0 50 100   200   300   400   500   600   700   800   900 1000 1100 1200   1350 m
                                                       b

Fig. 3-9. Map and profile of slush avalanche runout zone V45, Kärkevagge: June, 1958 (Site No. 2 in Fig. 11). (From Rapp, 1960.)

The snow avalanches at Mt. Nuolja, crossing the railway, are winter slab avalanches, as is clear from their form, content and time of occurrence. They vary in frequency from annual to 30-40 years recurrence intervals and occur in January-April. The largest cases dated in one track have the following time series: February 24, 1904; March 9, 1943; January 8, 1973.

Fig. 3-10. Slush avalanche track and deposit in front of Kärkevagge cirque, Kärkevagge. The avalanche boulder tongue is deposited upon an ulluvial fan (Site No. 3 in Fig. 11). (Photo A.R. 1980.)

In contrast, slush avalanches occur much later and mark the opening of the spring flood in their stream course. As an example we may mention that the slush avalanches at Mt. Kaisepakte in track 16 (figure 3-11) occurred on the following dates until protection was installed after 1940: May, 1916; May 25, 1929; May 28, 1934; May 18, 1938; June 2, 1940.

Other sites where very powerful slush avalanches have damaged structures are on the road near Suorva, Northern Sweden, where two tracks follow southfacing slopes with steep stream courses. In 1975, when these sites were monitored by S. Larsson (1975) they were triggered on separate days in connection with two situations of rapid snow melt (May 11 and May 22). At Longyearbyen, Spitsbergen, the worst case of destruction by a slush avalanche occurred on June 11, 1953, the so-called Haugen disaster, described by A. Jahn (1967) and others.

As regards damage done to human life and property, both debris flows and slush avalanches have caused considerable losses. Both from a theoretical point of view and for practical reasons of planning protection in connection with increasing tourism and construction in mountains, slush avalanches should be more thoroughly studied and better known. Several useful methods in such studies have been described in a recent literature review by Luckman (1977).

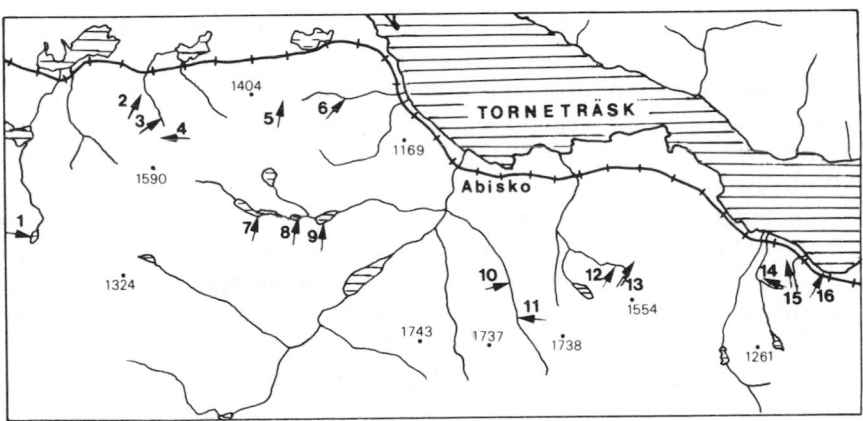

Fig. 3-11. Map of known sites of slush avalanches (arrows Nos. 1-16) in the Abisko area south of the railway. (From Nyberg, 1980.) Cf. Fig. 3.

CONCLUSIONS

Two different types of rapid mass movements in northern mountains have been described and discussed in this paper, namely, debris and slush avalanches. Debris flows that resulted from rainstorms of high intensity in summer or autumn have created major erosion and deposition on debris slopes, either in large areas, called regions, or in smaller areas, called cells or spots. Return periods of 50-200 years or longer seem likely for major debris flow events in the northern Scandinavian mountains above the forest limit. Flooding of downstream areas beyond the limits of hillslope failure has also occurred and caused increased erosion of channels and sedimentation in downstream lakes, with possible influence on aquatic life.

High intensity runoff during snowmelt creates favourable situations for slush avalanches. They have shorter return periods than debris flows and have a geomorphological importance localized to clearly identifiable tracks along steep first-order streams. Because of their comparatively frequent occurrences with a recurrence interval of 5-10 years, slush avalanches have a geomorphic significance of the same order of magnitude as the debris slides and flows. Slush avalanches and debris flows are two major types of rapid mass movements, the former with a frequency maximum in May and early June, the latter in late June, July and August.

The long return periods and low frequency of some of the processes in the periglacial areas is one reason for the necessity to make long-time studies and monitoring in order to understand the roles played by processes of low frequency/high magnitude in the sculpturing of mountain landforms.

Although this paper is based on observations during a rather long period, 1952 to 1980, it is in several respects only a progress report. We have not completed the field studies of the Nissunvagge case of alpine debris flows of 1979, to mention one ongoing study opening new doors. Other open questions are those of the internal and external geomorphic thresholds that control the distribution in time and space of slush avalanches in northern Scandinavia. These are follow-up studies from the work I did in Kärkevagge and elsewhere in the 1950s. It is now twenty years since I wrote in the preface of my thesis on Kärkevagge that ten years of study had not made me feel tired of the valley. I confess that almost thirty years of acquaintance with the problems of dynamical geomorphology in our northern study areas has not nearly exhausted our interest or stopped the flow of new and useful information.

REFERENCES

Coates, D.R. and Vitek, J.D., eds., 1980. Thresholds in geomorphology. G. Allen & Unwin, London: 498 pp.

Corner, G.D. 1980. Avalanche impact landforms in Troms, North Norway. Geog. Ann. *62A*: 1-10.

Curry, R.C. 1966. Observations of alpine mudflows in the Tenmile range, central Colorado. Geol. Soc. Amer., Bull. *77*: 771-776.

Fischer, A.G. 1969. Geological time-distance rates — the Bubnoff unit. Geol. Soc. Amer., Bull. *80*: 549-552.

Grove, J.M. 1972. The incidence of landslides, avalanches and floods in western Norway during the Little Ice Age. Arctic and Alpine Research. *4*: 131-138.

Jahn, A. 1967. Some features of mass movement on Spitsbergen slopes. Geog. Ann. *49A*: 213-225.

Karlén, W. 1973. Holocene glacier and climatic variations, Kebnekaise mountains, Swedish Lappland. Geog. Ann. *55A*: 29-62.

Larsson, S. 1975. Lavinrisken längs Ritsemvägen. Uppsalas Univ., Naturgeografiska Inst. Mimeo: 32 pp.

Luckman, B. 1977. The geomorphic activity of snow avalanches. Geog. Ann. *59A*: 31-48.

Nyberg, R. 1980. Slasklaviner i Abiskofjällen. Svensk Geog. Årsbok *56*: 47-56.

Rapp, A. 1959. Avalanche boulder tongues in Lappland. Geog. Ann. *41*: 34-48.

Rapp, A. 1960. Recent development of mountain slopes in Kärkevagge and surroundings, northern Scandinavia. Geog. Ann. *42*: 73-200.

Rapp, A. 1963. The debris slides at Ulvådal, western Norway, Nachr. Akad. Wiss. Göttingen, II. Math-Phys. Kl., Nr. 13: 195-210.

Rapp, A. 1974. Slope erosion due to extreme rainfall with examples from tropical and arctic mountains. Geomorphologische prozesse und Prozesskombinationen in der Gegenwart unter Verschiedenen Klimabedingungen. Nachr. Akad. Wiss. Göttingen, III. Math-Phys. Kl. Nr. *29*: 118-136.

Rapp, A. and Nyberg, R. 1981. Alpine debris flows triggered by a violent rainstorm on June 23, 1979, near Abisko, northern Sweden. 3rd Mtg. Int. Geog. Union Comm. on Field Experiments in Geomorphology, Kyoto, Japan, August, 1980. *In* Japan Geomorphological Union, Trans. *2*(2): 171-183.

Rapp, A. and Strömquist, L. 1979. Field experiments on mass movements in the Scandinavian mountains with special reference to Kärkevagge, Swedish Lappland. Studia Geomorphologica Carpatho-Balcanica *13*: 23-40.

Rudberg, S. 1950. Ett par fall av skred och ravinbildning i Västerbottens fjälltrafkter. Geol. Föreningens Stockholm Förhandlingar *72*: 139-148.

Sharpe, C.F.S. 1938. Landslides and related phenomena. N.Y., Columbia Univ. Press: 137 pp.

Starkel, L. 1976. The role of extreme (catastrophic) meteorological events in contemporary evolution of slopes. *In* Derbyshire, E., ed. Geomorphology and Climate. London, J. Wiley & Sons: 203-246.

Strömquist, L. 1976. Massrörelser initierade av extremnederbörd. Ett exempel från Andöya i Nordnorge. Norsk Geog. Tidsskrift *30*: 41-50.

Varnes, D.J. 1958. Landslide types and processes. Chapter 3 *in* Eckel, E., ed. Landslides and engineering practice. Highway Res. Bd., Washington, Spec. Rep. *29*: 20-47.

Wallén, C.C. 1959. The Karsa Glacier and its relation to the climate of the Torne Träsk region. Geog. Ann. *41*: 236-244.

Washburn, A.L. 1979. Geocryology: a survey of periglacial processes and environments. London, Edward Arnold: 406 pp.

Wolman, M.G. and Miller, J.P. 1960. Magnitude and frequency of forces in geomorphic processes. J. Geology *68*: 54-74.

Zaruba, Q. and Mencl, V. 1969. Landslides and their control. Prague, Academia and Amsterdam, Elsevier: 214 pp.

Zenzén, N. 1926. Några upplysningar rörande jordskredet på Ö. Stårbetjvare i Arjeplog, Pite Lappmark den 3 aug. 1920. Geol. Föreningens Stockholm Förhandlingar *48*: 167-185.

4

ESTIMATION OF AVALANCHE RUNOUT DISTANCES IN NEW ZEALAND

B. B. Fitzharris

Department of Geography, The University of Otago
Box 56, Dunedin New Zealand

ABSTRACT

Recreation and tourist pressures in the snow zones of New Zealand mountains are increasing the avalanche hazard. Zoning maps are needed for heavy use areas but limited historical records make zoning difficult. Furthermore, most avalanches run out onto grassland, so trimlines in trees that record past large events are rare. Four models for estimating maximum runout distance have been tested on five paths in the Mount Cook region where evidence of past avalanche activity is available. The average slopes of paths measured from fracture line to maximum runout distance are similar to those found in Norway. Voellmy coefficients for sliding friction (μ) are similar to those used in the Northern Hemisphere. Those for turbulent friction (ξ) are: lower (150-650 m s^{-1}), which probably reflects the rough and generally snowfree nature of many New Zealand runout zones.

RÉSUMÉ
Évaluation des distances de parcours des avalanches en Nouvelle-Zélande

Les activités touristiques et récréatives dans les régions de niege des montagnes de la Nouvelle-Zélande accroissent les dangers liés aux avalanches. Des cartes de zonation sont requises pour les régions d'usage intensif, mais la documentation historique limitée' rend cette tache difficile. De plus, la plupart des avalanches débouchent sur la prairie, donc les cicatrices de forêts qui délimitent des évenements majeurs sont rares. Quatre modèles d'évaluation de la distance maximale de parcours sont évalués à partir de cinq couloirs d'avalanches près du Mont Cook où les traces d'avalanches anciennes sont préservées. Les pentes moyennes des avalanches depuis la zone de fracture jusqu'à la limite extrême du parcours sont semblable à celles mésurées en Norvège. Les coefficients de Voellmy pour la friction de glissement (μ) sont semblables à ceux utilisés dans l'hémisphère nord. Ceux pour la friction de turbulence (ξ) sont inférieurs (150 - 650 ms^{-1}), ce qui réflète probablement la nature rugueuse et généralement libre de neige de beaucoup de zones de parcours d'avalanche en Nouvelle-Zélande.

ОЦЕНКА РАССТОЯНИЯ ВЫБЕГА ЛАВИН В НОВОЙ ЗЕЛАНДИИ

Б. Б. ФИТСХАРРИС

РЕЗЮМЕ

Присутствие туристов и отдыхающих в снежной зоне Новой Зеландии повышает опасность снежных лавин. Нужны зональные карты территорий интенсивного пользования, но недостаток исторических данных затрудняет зонирование. К тому же, большиство лавин выбрасываются на лугопастбищные угодья и поэтому линии подрезки на деревьях, которая служит показателем важных событий в прошлом, встречается сравнительно редко. Проводятся испытания четырех моделей, оценивающих максимальные расстояния выбега лавин на пяти трассах в районе Маунт Кук, где существуют информационные данные относительно лавинной деятельности в прошлом. Средний наклон падения, измеряемый от линии излома до максимального расстояния выбега, аналогичны встречаемым в Норвегии. Коэффициенты трения скольжения Воэллмы ( ) тоже аналогичны применяемым на северном полушарии. Коэффициенты турбулентного трения ( ) ниже (150-650 мс$_{-1}$), что, вероятно, указывает на шероховатые и свободные от снега зоны выбега многих лавин в Новой Зеландии.

INTRODUCTION

New Zealand is a mountainous country, yet almost all settlement is on coastal lowlands or in valleys at elevations below 700 m and few public roads attain heights above 1000 m. Because of the country's maritime location snow seldom descends to the coast and, even in the south, the winter snowline normally remains above 1000 m. Most economic activity takes place below this elevation, so that New Zealanders are a lowland people living in the snow-free zone of a mountainous land.

However, the mountains provide an attractive recreational resource and, with the escalation of mountain leisure activities, the hazard from snow avalanches is rising (Dingwall, 1977; La Chapelle, 1979). The extensive development of ski and tourist areas now brings thousands of people and numerous buildings into the avalanche zone for the first time. Tourist traffic on avalanche prone roads is also increasing. Several recent fatalities and damage to ski buildings and structures suggest that there is a need for avalanche hazard zoning maps for popular areas.

It is common practice in North America to determine maximum runout distances by inspecting damage to trees. Clearly defined patterns of tree species are sometimes present in and around avalanche tracks and such evidence is often the basis of avalanche zoning maps (Perla & Martinelli, 1976). Unfortunately, many avalanche prone areas in New Zealand are colonized by grasses and low scrub, which do not record clearly the history of past avalanche events. Furthermore, because European colonization began as recently as the nineteenth century and few people live in the mountains, there is a dearth of historical documentation of avalanches. As a consequence, it is difficult to find sufficient evidence to produce reliable zoning maps for many areas of New Zealand.

One method of zoning is to model the runout distance for large, infrequent avalanches (the maximum runout distance). Such models have not been tested in New Zealand, but for much of its mountainous area, this is the only method available for preparing a zoning map. Therefore, the aim of this paper is to select a range of models and to apply them in the Mount Cook region. In this way, the most effective model can be chosen for avalanche zoning. Five avalanche paths near Mount Cook village have been selected for testing (figure 4-1). Here, rare, taller vegetation sometimes allows definition of trimlines, and anecdotal and recorded observations of avalanche activity are available to estimate maximum runout distance. As well, a zoning map is required to plan future development at this popular tourist area. All except the Ollivier path have channelled tracks. Black Birch runs out along a narrow valley, but the other four paths spill onto rocky, partially forested debris fans that are normally free of winter snow.

TOPOGRAPHY, CLIMATE AND VEGETATION

The Mount Cook region lies immediately to the east of the highest part of the Main Divide of the Southern Alps. Mount Cook is the main peak (3764 m), with many other mountains above 3000 m. A series of lower ranges, such as the Mount Cook Range and the Sealy Range, run southwest to northeast within the region. They are separated by the deep, glaciated valleys of the Hooker, Mueller and Tasman glaciers. The main rock types are Alpine schists in the west and lower grade metamorphics and greywackes towards the east.

The avalanche paths chosen for this study (see figure 4-1) have their start in cirque basins of the seasonal snow zone (1000-2000 m). Many paths run down gullies and across large debris fans that adjoin the valley floor (eleva-

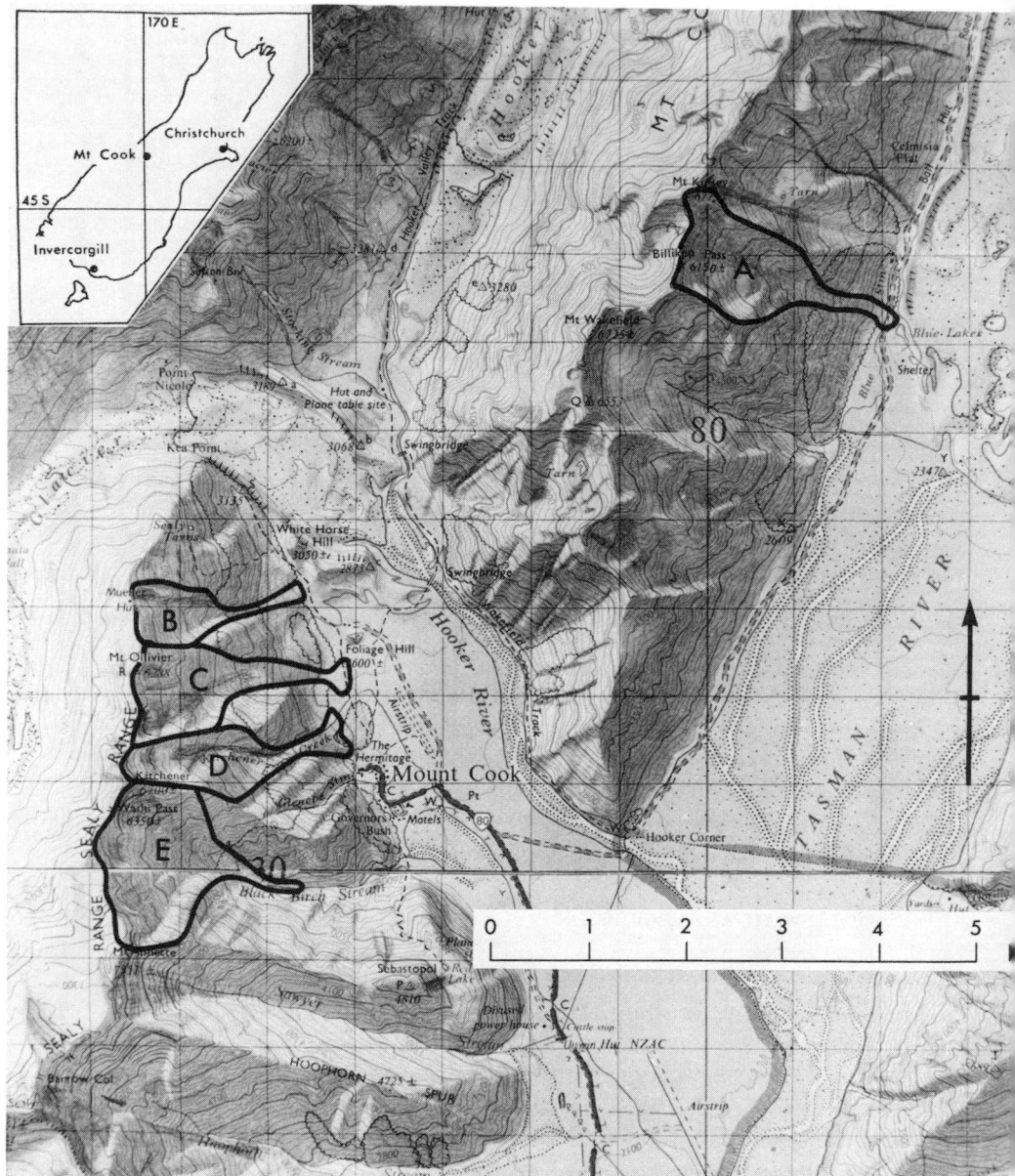

Fig. 4-1. The Mount Cook Village region, showing the five avalanche paths studied: A = Billiken Pass, B = White Horse, C = Ollivier, D = Kitchener, E = Black Birch. Inset shows study area location in the South Island of New Zealand.

Fig. 4-2. The Ollivier avalanche path, illustrating how avalanches usually run out below the normal winter snowline onto a well vegetated fan (courtesy of Mount Cook National Park).

tion 600-800 m: figure 4-2). The runout zones are usually free of snow, even in winter. Avalanches are also common in the perennial snow zone (above 2000 m), but are not considered here.

Annual precipitation at Mount Cook village (elevation 765 m) averages 4071 mm, but probably exceeds 10,000 mm near the Main Divide (Chinn, 1979). Precipitation is lower in winter (July average 244 mm) than in summer (January average 401 mm). Temperatures average 1.6°C in July but have ranged between extremes of -12.8°C and 17.8°C. In general, the atmospheric freezing level intersects the mountains at some elevation between valley floor and mountain tops, but can fluctuate widely even within a day. Winter and spring snowlines usually intercept the mountains above the valley floor. Snow accumulation increases rapidly, often at a rate greater than 0.5 m water equivalent per 100 m increase in elevation (Fitzharris, 1978). New Zealand mountains are exposed to the Southern Westerlies, and hence are subject to persistently high winds. Both the fluctuating freezing levels and wind drifted snow influence local avalanche formation (Fitzharris, 1976).

There is little vegetation in the avalanche starting zones except for limited areas of alpine grassland, with occasional scrub below 1300 m. The dominant species of the starting zone and track elevation are snow tussock (*Chionochloa sp*), shorter grasses (e.g. *Poa colensoi*), turpentine (*Dracophyllum sp*), snow totara (*Podocarpus nivalis*), shrub daisies (*Olearia sp.; Cassinia sp*) and *Hebe*. Lower runout zones display similar scrub species, but can have taller forest vegetation, such as totara (*Podocarpus hallii*), broadleaf (*Griselinia littoralis*) and silver beech (*Nothofagus menziesii*). Vegetation in active parts of the avalanche paths often gives way to talus, although a few specialised plants such as avalanche grass (*Poa cockayniana*) have adapted to impact and can survive delayed snow melt.

THE MODELS

Four models were chosen for testing:

1. Norwegian model
Lied and Bakkehøi (1980) defined the average gradient of an avalanche path as

$\phi = \arctan (Y/X)$

where   Y = vertical height difference between the highest point of rupture in the starting zone and the outer end of avalanche debris in the runout zone (m).

X = horizontal displacement between these two points (m).

On the basis of the detailed registration of 423 Norwegian avalanches over more than 100 years, they found that ϕ at the maximum runout distance varies between 18° and 50°. However, where the starting zone is a cirque, ϕ is equal to or less than 25° for many avalanche paths.

Using this rule of thumb approach, an estimate of maximum runout distance may be found by moving down the slope of an avalanche path until $\phi = 25°$.

2. Regression model

Bovis and Mears (1976) developed a least squares regression equation for Colorado,

$$S = 213.7 + 11.4\,A$$

where   S = maximum runout distance (m), measured from the beginning of the runout zone, here taken as the point where the slope equals 16°;

A = area of starting zone (hectares).

The correlation coefficient was 0.81.

They argued that such a relationship was reasonable because the starting zone area controls the size of slab likely to fracture during a large avalanche. The size of slab release determines the discharge rate of snow to the avalanche track and, therefore, the hydraulic radius. This in turn controls the velocity of the avalanche in the track, assuming that flowing snow may be analysed by conventional fluid mechanics methods.

3. Voellmy model

Voellmy (1955) was the first to modify principles of classical fluid dynamics for avalanche problems. His equations have since been improved by others (e.g. Schaerer, 1973, 1975; Mears, 1976), but the underlying concepts remain unchanged. Only the basic equations are given here, but a more detailed exposition was given by Leaf and Martinelli (1977). They demonstrated that reasonable values for S can be obtained with proper field data and equation coefficients.

Voellmy's expression for runout distance is

$$S = v^2/[2g(\mu \cos \theta - \tan \theta) + v^2 g/\xi E_m] \tag{2}$$

where   g = acceleration due to gravity (9.8 m s^{-2});

μ = coefficient of sliding friction;

θ = slope of avalanche path (degrees);

v = velocity (m s^{-1});

ξ = coefficient of turbulent friction (m s^{-2}).

When the debris is piled in a short steep cone

$$E_m = h' + v^2/4g \tag{3}$$

where   h' = vertical height of the avalanche flow (m).

The velocity at the start of the runout zone is assumed to be the terminal velocity, as given by

$$v^2 = \xi h'(1 - \gamma a/\gamma s)(\sin \theta - \mu \cos \theta) \tag{4}$$

where γ_a = specific weight of air (kg m^{-3})

γ_s = specific weight of flowing snow (kg m^{-3}).

The velocity is assumed to diminish uniformly to zero, lower in the runout zone. In confined channels, h' is replaced by R, the hydraulic radius:

$$R = a/P$$

where -a = cross sectional area (m^2)

P = "wetted" perimeter (m).

Schaerer (1975) has suggested

$$\mu = W/v \leqslant 0.5 \tag{5}$$

where W = a parameter which he found to be 5 m s^{-1}. Schaerer also pointed out that ξ is dependent on the condition of the avalanche path, varying from 150 m s^{-2} for slopes with boulders, trees and forests, to 1800 m s^{-2} for those with a smooth snow cover. For flowing and mixed motion avalanches, Leaf and Martinelli (1977) reduced equation (4) to

$$v^2 = \xi[R \sin \theta - \mu d \cos \theta] \tag{6}$$

where d = a/w is hydraulic depth (m),

and w = top width of the cross-section (m).

The Voellmy theory is recognized as being somewhat simplistic, but Moskalev (1966) showed that equations that are theoretically more rigorous tend to be too complex and offer little practical advantage.

4. Numerical model

Perla et al. (1980) have described a numerical model for computing v as a function of S on a slope of given geometry. v(S) is computed from the differential equation of motion

$$dv^2/dS = 2 g(\sin \theta - \mu \cos \theta) - 2 Jv^2/M \tag{7}$$

where M = avalanche mass (kg)

J = drag parameter (kg m^{-1}).

The parameter $M/J = \xi h'/g$, as used in the Voellmy approach, and is the mass to drag ratio.

Equation (7) is solved numerically, as discussed by Cheng and Perla (1979), by dividing the slope into segments small enough so that θ can be considered constant over the length (L) of the segment. Each segment is assigned values for θ, L, μ, and M/J. If the speed at the beginning of segment n is v_n, and the avalanche does not stop, then the speed at the end of the segment (beginning of segment n+1) is a solution of equation (7):

$$v_{n+1} = \sqrt{[\alpha_n(M/J)_n (1 - \exp \beta_n) + v_n^2 \exp \beta_n]} \tag{8}$$

where $\alpha_n = g(\sin \theta_n - \mu_n \cos \theta_n)$;

$\beta_n = -2L_n/(M/J)_n$

If the avalanche stops within the segment, the runout distance from the beginning of the nth segment is the equation (7) solution

$$2S = (M/J)_n \ln [1 - v_n^2/\alpha_n(M/J)_n] \qquad (9)$$

The computation is iterated down the runout zone until the stopping position is reached.

While similar to the Voellmy model, the numerical approach allows flow parameters to be varied at increments along the path, thus adding flexibility to the model.

An alternative numerical approach using a finite difference computer programme based on the Navier-Stokes equations has been produced by Lang *et al*. (1979).

METHODS

The assessment of maximum runout distance was made from anecdotal evidence of "old timers," experienced climbers, and staff of the Mount Cook National Park. The Park has recorded avalanche activity since 1975 and has archived earlier records, including photographs of debris, dating back 30 years (figures 4-2 and 4-3). Avalanche activity on the Kitchener,

Fig. 4-3. A mixed motion avalanche on the Ollivier path, 13 August 1980. The photograph is taken looking west from Mount Cook Village. The subsequent avalanche debris is shown in Figure 4-2 (courtesy of Mount Cook National Park).

Ollivier and White Horse paths can be observed readily from popular walking tracks and roads close to Mount Cook Village. The Ball Hut road is sufficiently used for major avalanche activity on the Billiken Pass path to be always noted. Black Birch cannot be seen from the Village area but is often visited by walkers or Park Rangers.

Vegetation about the track and runout zone was inspected in the field and on aerial photographs to identify patterns of damage and species distribution, as discussed for New Zealand conditions by Burrows *et al.* (1979). This evidence provides estimates of maximum runout distance, and sometimes flow height, h', for at least the last 100 years.

Values of ϕ were measured using Abney level sightings from the maximum runout distance. Starting zones were defined as in Bovis and Mears (1976) from aerial photographs, field inspection, and historical records of actual fracture lines (Mount Cook National Park Board). Areas of starting zones were obtained from maps (scale 1:25,000).

Slope angles required for the Voellmy and for the numerical models were

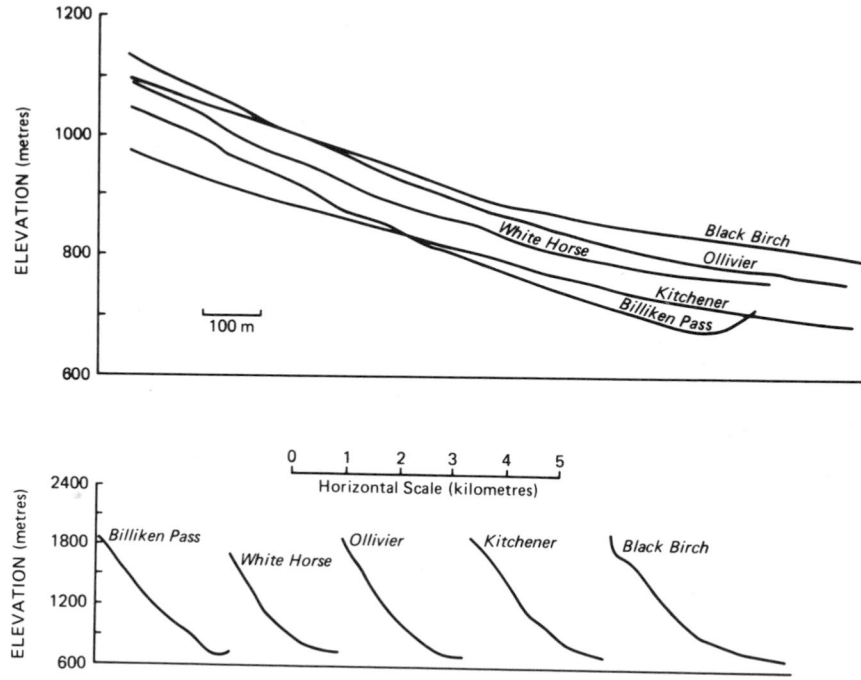

Fig. 4-4. (a) Profiles of avalanche paths.
(b) Detailed profiles of lower track and runout zone as surveyed by theodolite.

measured over the lower portion of the paths by theodolite surveys (figure 4-4b). Traverses covered all of the runout zone and lower portions of the track. On debris fans, where avalanches can travel a number of routes, several transects were taken. Slopes of the upper track and starting zone were estimated from maps and from Abney sightings in the field (figure 4-4a).

A computer programme was written to calculate the Voellmy runout distance using equations (6) and (2), and operated iteratively using various values of μ and ξ until the maximum runout distance was obtained. Where h' could not be estimated from field evidence, it was computed using methods reviewed by Leaf and Martinelli (1979, p 7-9). The computer programme of Cheng and Perla (1979) was similarly iterated for various combinations of μ and M/J until the simulated avalanche stopped at the maximum runout distance. These procedures do not provide a unique solution for μ, ξ, or M/J but a range of options.

RESULTS

Estimates of maximum runout distance are expressed in Table 4-1 as the elevation to which an avalanche can run. Large events observed in the past involved slab fracture 1 to 2.5 m deep and mixed motion avalanches, both of the direct action and climax types (Fitzharris, 1976; Ho, 1982; see example in Figure 4-3). Dry powder avalanches, which could run further, have not been observed on the paths considered here, but occur at higher elevations

TABLE 4-1

ESTIMATED ELEVATION OF THE MAXIMUM RUNOUT LIMIT
AND VALUES OF ϕ OBTAINED FROM THE NORWEGIAN MODEL

Path	Elevation m (and feet)	Evidence	ϕ (degrees)
Billiken Pass	730 (2400)*	Old photographs; hearsay; damaged vegetation	25
Kitchener	760 (2500)	Vegetation patterns and damage	26
Ollivier	730 (2400)	Vegetation patterns; damage; photographs; National Park Board records	26
White Horse	780 (2550)	Vegetation patterns	27
Black Birch	850 (2800)	Vegetation patterns (well defined trimline); National Park Board records; hearsay	24

* *adverse slope.*

within Mount Cook National Park. From the historical data available at Mount Cook and that from the Milford Road to the southwest (see Fitzharris and Owens, 1980), big avalanches appear to have occurred in 1936, 1945, 1946, 1957, 1964, 1968, 1972, 1980 and 1982.

Avalanches reach their lowest elevation at Billiken Pass, where the largest run into Blue stream, across Ball Hut road and up the adverse slope formed by the lateral moraine of Tasman Glacier. At Kitchener, Ollivier, and White Horse they run onto the debris fan, but do not reach the outwash of the valley floor (cf figure 4-2). At Black Birch, they fail to reach the fan. Longitudinal profiles of the paths display concave slopes (figure 4-4), but are remarkably even in the runout zone.

1. Norwegian model

Values of ϕ for the five paths are similar to the 25° reported from Norway (Table 4-1). Considering the "rule of thumb" nature of this method, no significance can be attached to the variations observed in ϕ. A variation of ±1° in ϕ produces ±200 m variation in runout distance and ±40 m in Y on Black Birch. Other paths show smaller variation. In Norway, ϕ can approach 18°, given large, deep slab fractures, smooth tracks, and snow covered runout zones. It seems unlikely that ϕ could reach 18° on the Mount Cook paths, as the lower slopes of runout zones are rough, heavily vegetated, and snow depth is never sufficient to make the path smooth.

2. Regression model

This provided variable results (Table 4-2). When using this method there are difficulties in deciding the area of the starting zone and the beginning of the runout zone. Paths such as Billiken Pass and Black Birch have distinct multiple starting zones, so that each is added to find A. This procedure can lead to large values of S, which appear to be unrealistic because it is unlikely that all of the multiple starting zones would fracture simultaneous-

TABLE 4-2

ESTIMATED MAXIMUM RUNOUT DISTANCE AND VALUES COMPUTED USING THE REGRESSION MODEL OF BOVIS AND MEARS (1976)

Path	Starting Zone Area $(m^2 \times 10^4)$	Observed maximum runout distance (m)	Model maximum runout distance (m)
Billiken Pass	43	500	700
Kitchener	22	520	465
Ollivier	27	400	520
White Horse	15	420	385
Black Birch	52	240	805

ly. An empirical model such as that of Bovis and Mears (1976) cannot be expected to be successful in a topographic and climatic environment different from that in which it was formulated.

3. Voellmy model

Values of μ and ξ which gave the best fits are summarised in Table 4-3. Generally, μ varies between 0.20 and 0.30, and ξ between 150 and 650 m s⁻². These values for ϕ are low compared with those for Northern Hemisphere paths. For example, in North America Leaf and Martinelli (1977) found $500 < \xi < 1800$ m s⁻². Many authors consider that, for zoning purposes, relatively high values of ξ should be used to simulate major events. However, the runout zones of the Mt. Cook paths appear to be rougher and more snow-free than those tested in the Northern Hemisphere. Values of ξ between 150 and 650 m s⁻² are consistent with the findings of Schaerer (1975), who suggested the following (in m s⁻²):

smooth cover, no trees	1200 - 1800
average, open mountain slope	500 - 750
average gully	400 - 600
slope with boulders, trees, forests	150 - 300

Values of μ are similar to those of Leaf and Martinelli (1977), who note that for normal snow conditions, μ varies from 0.15 to 0.20 in the upper part of the runout zone, and up to 0.50 at the end. For an avalanche to slow and stop, $\mu > \tan \theta$. Values of μ in Table 4-3 meet this criterion.

Overall, the best fit values of μ and ξ in Table 4-3 indicate mixed motion avalanches flowing over rough terrain. When $\mu \rightarrow 0$, and ξ becomes large, simulated avalanches run substantial distances beyond the observed max-

TABLE 4-3

RANGE OF BEST FIT VALUES FOR μ AND ξ
AS USED IN THE VOELLMY MODEL

Path	μ	$\xi(m\ s^{-2})$
Billiken Pass	0.25	150
	0.30	500
Kitchener	0.30	150
	0.40	500
Ollivier	0.20	200
	0.30	650
White Horse	0.25	150
	0.30	500
Black Birch	0.20	150
	0.30	500

imum runout distances. However, this condition, which represents powder avalanches flowing over a small, deeply snow covered runout zone, is not appropriate for the paths studied here. In four out of nine cases analysed by Martinelli *et al* (1980), ξ varied between 700 and 800 m s^{-2}, with the others ranged between 1200 and 2000 m s^{-2}. Buser and Frutiger (1980) recommended that to model extreme flowing avalanches in Switzerland, $\mu = 0.16$ and $\xi = 1360$ m s^{-2} be used, although standard practice has been to choose $\xi = 500$ m s^{-2}.

4. Numerical model

A range of values for μ and M/J provide simulated avalanches which stop at the maximum runout distance (Table 4-4). μ varies over a similar range to that in Table 4-3, with M/J varying from 50 to 800 m. Corresponding values for ξ (Table 4-3) are in broad agreement where h' is known. For example, trim line evidence at Black Birch suggests that for large avalanches, h' = 14 m, which gives $\xi = 175$ for $\mu = 0.20$, and $\xi = 350$ for $\mu = 0.25$, values that are consistent with Table 4-3.

Bakkehøi *et al* (1981) noted that solutions to equation (7) are insensitive to variations in M/J. They showed that the model predicts runout distances that are not very different from each other when markedly different μ and

TABLE 4-4

RANGE OF BEST FIT VALUES FOR μ AND M/J
AS USED IN THE NUMERICAL MODEL

Path	μ	M/J (m)
Billiken Pass	0.10	100
	0.20	250
	0.25	400
	0.30	500
	0.35	700
Kitchener	0.25	50
	0.30	500
	0.35	800
Ollivier	0.20	100
	0.25	250
	0.30	500
White Horse	0.15	50
	0.20	150
	0.25	400
	0.25	550
Black Birch	0.15	50
	0.20	250
	0.25	500

M/J are used as input. This insensitivity of the model is displayed in Table 4-4 (and by implication in the Voellmy model results in Table 4-3), where there are radical variations in M/J (and ξ) as compared with the range of μ.

Perla *et al* (1980) found best fits of M/J from 125 to 2500 for 25 Cascade highway avalanches in the United States. These paths have been re-examined by Bakkehoi *et al* (1981), together with 111 from Norway. They computed pairs of μ and M/J that matched known boundaries, and then excluded those which were inconsistent with the few published data describing v. Best results were found by setting M/J in the approximate range 10 Y > M/J > 0.1Y, and setting μ in the approximate range 0.5 > μ > 0.1. For μ = 0.20, Mount Cook data yield $\overline{(M/J)}$ = 0.2 $\overline{Y}$, and for μ = 0.25, $\overline{(M/J)}$ = 0.3 $\overline{Y}$. These results tend to be in the lower range of Bakkehøi *et al*, who note that low values of M/J are applicable to slow moving avalanches which consist of wet snow. Thus, the Mount Cook values are consistent with flowing or mixed motion avalanches moving over rough terrain.

CONCLUSIONS

Of the four models tested, all except the regression approach offer promise for predicting maximum runout distance for New Zealand avalanches. The simply applied Norwegian method gave surprisingly good results, with $\phi \approx$ 24-27°. The Voellmy and the numerical models also provided tractable results, but the coefficients μ, ξ and M/J must be chosen appropriately from a range of values. Suitable values of ξ between 150 and 650 m s^{-2} are less than those found in the Northern Hemisphere, probably because the Mount Cook runout zones are rougher and generally snow free in winter, and avalanches there tend to be of the flowing type.

The starting zones of the five Mount Cook paths have similar aspects, elevations and climatic regimes, so coefficients could be different elsewhere in New Zealand. Therefore data for other paths are needed to refine these estimates regionally.. Runout zones farther north in the Craigieburn Range marked by well-defined vegetation patterns are currently being surveyed. Avalanche records back to 1936 are available near Milford Sound (Fitzharris and Owens, 1980), but most paths are cliff-like and not easily handled by the models. Compared with historical data available from European countries, the record in New Zealand is short. Hence, the possibility exists that future avalanches at Mount Cook could exceed the maximum runout distances listed in Table 4-2.

ACKNOWLEDGEMENTS

I wish to thank R. Clark, A. Davidson, K. Day, K. Paulin, M. Sutherland and K. Williams, who surveyed the avalanche paths, and the Mount Cook National Park Board, which supplied avalanche information.

REFERENCES

Bakkehøi, S., Cheng, T., Domaas, U., Lied, K., Perla, R. and Schieldrop, B. 1981. On the computation of parameters that model snow avalanche motion. Can. Geotech. J. *18*: 121-130.

Bovis, M.J. and Mears, A.I. 1976. Statistical prediction of snow avalanche runout from terrain variables in Colorado. Arctic and Alpine Research, *8*: 115-120.

Burrows, C.J., Conway, B.R. and McIntosh, T. 1979. Snow avalanches and vegetation. In Snow Avalanches, a review with special reference to New Zealand. New Zealand Mountain Safety Council, Avalanche Comm. Rep. *1*: 56-73.

Buser, O. and Frutiger, H. 1980. Observed maximum runout distance of snow avalanches and the determination of the friction coefficients μ and ξ. J. Glaciology, *26*: 121-130.

Cheng, T.T. and Perla, R. 1979. Numerical computation of avalanche motion. Canada Dept. Environment, Inland Waters Directorate, National Hydrology Research Institute, NHRI Paper *5*: 12 pp.

Chinn, T.J. 1979. How wet is the wettest of the wet West Coast? N.Z. Alpine J. *32*: 85-87.

Dingwall, P. 1977. The avalanche hazard in New Zealand. N.Z. Dept. Lands and Survey, Information Series No. 2: 33 pp.

Fitzharris, B.B. 1976. An avalanche event in the seasonal snow zone of the Mount Cook Region, New Zealand. N.Z. J. Geology and Geophysics *19*: 449-62.

Fitzharris, B.B. 1978. Problems in estimating snow accumulation with elevation on New Zealand mountains. J. Hydrology (N.Z.) *17*: 78-90.

Fitzharris, B.B. and Owens, I.F. 1980. Avalanche atlas of the Milford Road and an assessment of the hazard to traffic. N.Z. Mountain Safety Council, Avalanche Comm. Rep. *4*: 79 pp.

Ho, C.W. 1982. Snow avalanche studies in the Mount Cook region, M.Sc. Thesis, The University of Otago, Dunedin: 323 pp.

La Chapelle, E.R. 1979. An assessment of avalanche problems in New Zealand. N.Z. Mountain Safety Council, Avalanche Comm. Rep. *2*: 53 pp.

Lang, T.E., Dawson, K.L., and Martinelli, M. 1979. Numerical simulation of snow avalanche flow. U.S. Dept. Agriculture, Forest Service, Rocky Mountain Forest and Range Exp. Stn., Research Paper *RM-205*: 51 pp.

Leaf, C.F. and Martinelli, M. 1977. Avalanche dynamics: engineering applications for land use planning. U.S. Dept. Agriculture, Forest Service, Rocky Mountain Forest and Range Exp. Stn., Research Paper *RM-183*: 51 pp.

Lied, K. and Bakkehøi, S. 1980. Empirical calculations of snow avalanche runout distance based on topographic parameters. J. Glaciology *26*: 165-178.

Martinelli, M., Lang, T.E., and Mears, A.I. 1980. Calculations of avalanche friction coefficients from field data. J. Glaciology *26*: 109-19.

Mears, A.I. 1976. Guidelines and methods for detailed snow avalanche hazard investigations in Colorado. Colorado Geol. Surv., Bull. *38*: 125 pp.

Moskalev, Yu. D. 1966. Avalanche mechanics. Hydrometeorological Publishing House, Leningrad. (tr. U.S. Army Corps of Engineers Cold Regions Research and Eng. Lab., Hanover, N.H. 1970. NTIS: AD71 5049: 152 pp.)

Perla, R. and Martinelli, M. 1976. Avalanche handbook. U.S. Dept. Agriculture, Forest Service, Agriculture Handbook *489*: 238 pp.

Perla, R., Cheng, T.T., and McClung, D.M. 1980. A two parameter model of snow avalanche motion. J. Glaciology *26*: 197-208.

Schaerer, P.A. 1973. Observations of avalanche impact pressures. *In* Advances in North American Avalanche Technology. U.S. Dept. Agriculture, Forest Service, Rocky Mountain Forest and Range Exp. Stn., Tech. Rep. *RM-3*: 51-54.

Schaerer, P.A. 1975. Friction coefficients and speed of flowing avalanches. Int. Assoc. Sci. Hydrology *114*: 425-432.

Voellmy, A. 1955. Uber die Zerstörungskraft von Lawinen. Schweiz. Bauzeitung *73*: 159-165; 212-217; 246-249; 280-285. Tr. On the destructive force of avalanches. U.S. Dept. Agriculture, Forest Service, Alta Avalanche Study Center, Transl. 2 (1964): 63 pp. (Available from Rocky Mountain Forest and Range Expt. Stn., Fort Collins, Colo.).

5

THE ICE FACTOR IN FROZEN GROUND

L. W. Gold

Division of Building Research, The National Research Council of Canada, Ottawa, Ontario K1A 0R6 Canada

ABSTRACT

The nature of ice formations in frozen ground has been a major subject of investigation for Ross Mackay. In this paper, characteristics of the freezing of water in porous materials are reviewed and examples given of the effects of frost action. The thermodynamic basis for the explanation of ice segregation is given and the possible role of spreading and disjoining pressures is described. A criterion for the initiation of ice lenses is presented. Brief consideration is given to the question of whether equilibrium can ever be attained under natural conditions.

RÉSUMÉ
Le facteur glace dans le sol congélé

La nature des formations de glace dans le sol congélé est un thème de recherche majeur de Ross Mackay. Dans cet article, les caractéristiques de l'engel de l'eau dans des matériaux poreux sont précisés et des exemples sont donnés des effets de l'action du gel. La base thermodynamique pour expliquer la ségrégation de la glace est donnée et le rôle possible des pressions de dispersion et de disjonction est décrit. Une critèrium d'initiation des lentilles de glace est présenté. La question de savoir si l'équilibre est jamais atteint dans des conditions naturelles est brièvement examinée.

ФАКТОР ЛЬДИСТОСТИ В МЕРЗЛОМ ГРУНТЕ

Л. В. ГОЛД

РЕЗЮМЕ

Характер льдообразования в мерзлом грунте является главным предметом исследований проводимых Росс Макаем. В настоящем докладе рассматриваются характеристики замерзания воды в пористых породах и даются примеры эффекта морозобойного действия. Представлены термодинамические основания для выяснения сегрегации льда и описывается возможная роль распространяющегося и разъединяющего давлений. Дается критерий для зарождения ледяных линз. Коротко рассматривается возможность достижения устойчивого равновесия при обычных природных условиях.

INTRODUCTION

The forms of ice that occur in the ground can vary from massive bodies to minute inclusions; the average weight of ice in permafrost several metres thick can be more than five times that of the dry soil (Mackay, 1971). Ross Mackay has devoted considerable effort to delineating the various types of ground ice and to determining their origin and mode of formation (Mackay, 1972). From his studies of our northern regions he has added greatly to knowledge of ice formed in horizontal layers or lenses by forces associated with the freezing of water in some porous materials, a process called ice segregation or frost action (Mackay, 1972); of ice formed in permafrost due to the filling of cracks with water from the surface, for example, ice wedges (Mackay, 1974); of ice formed by the intrusion and subsequent freezing of water with or without segregation, for example, pingos (Mackay, 1973, 1979); of homogeneously distributed pore ice formed by water freezing *in situ* (Mackay, 1972); of ice crystals formed in underground cavities by sublimation (Mackay, 1972); or buried ice resulting from geological activity (Mackay, 1972).

These are some of the geologic manifestations of ground ice that Ross Mackay has investigated. There has been a growing appreciation of the importance of the insight and knowledge he has provided through these investigations. The performance of roads, buildings and other structures in the North has demonstrated the need to know the geological structure and thermal condition of the ground on which they are placed. The presence of ice, its form, its distribution and how it will respond to disturbance of the surface and to loads must be taken into consideration in the design, construction and operation of these structures. If one wants an outstanding example of contributions of scientific activity to engineering and economic development, one need only look at the use made of Ross Mackay's work in current northern oil and gas development.

For several years scientists and engineers have had a very active interest in ice segregation or frost action because of the damage it can cause to roads, buildings and other structures. This phenomenon is of great concern at present because of the severe conditions it poses for the operation of gas pipelines in the North. It is also one that interests Ross Mackay, with respect both to its possible role in the formation of ground ice and to its effects during the freezing of the active layer.

Frost action in soils and building materials has been an important subject of research at the Division of Building Research, National Research Council of Canada, for several years. The effects of frost action and some of its characteristics, as demonstrated by this work and that of others, are briefly reviewed in this paper. This is followed by a statement of the thermodynamic basis for the ice segregation process. The concepts of spreading and disjoining pressures in liquid films are discussed with respect to their possible role in frost action, and brief consideration is given to the effects of gradients of temperature and pressure on the stability of ground ice.

EFFECTS OF FROST ACTION

Perhaps the most common evidence of frost action is the heaving that occurs in some road surfaces in winter. Such heaving demonstrates one of the characteristics of frost action; that is, it is a segregation of ice that causes an increase in volume greater than can be attributed to the expansion of water on freezing. In fact, the process involved does not depend on that property of water; a similar behaviour can be obtained with other liquids.

In his classical work published in 1935, Beskow stated that observations on ground ice and frost action were recorded as early as 1700. Both he and Tabor (1930) stated that not until the early 1900s was it recognized that during the freezing of fine-grain soils water can flow to the freezing zone and that the phenomenon of heaving is due to ice segregation rather than to expansion of water on freezing.

Pressures developed by frost action can be quite large. Damage to improperly protected foundations on frost susceptible soils is not uncommon. A total upward force of about 130 kN (30,000 lb) was measured by Penner (1970) on a 30 cm diameter plate placed on clay ground and prevented from heaving upwards. This behaviour is not confined to soils. Figure 5-1(a) shows displacement of brick and coping due to the formation

Fig. 5-1. (a) Damage to coping due to frost action.
(b) Damage to brick due to formation of ice lenses.

of ice, and figure 5-1(b) shows the splitting of brick considered to be due to ice lens formation.

Research on frost action soon demonstrated that the tendency to heave during freezing depends upon the characteristics of the soil, particularly grain size, the rate of heat extraction, pressure in the direction of freezing, water content and the pressure or tension in the soil water. Figure 5-2

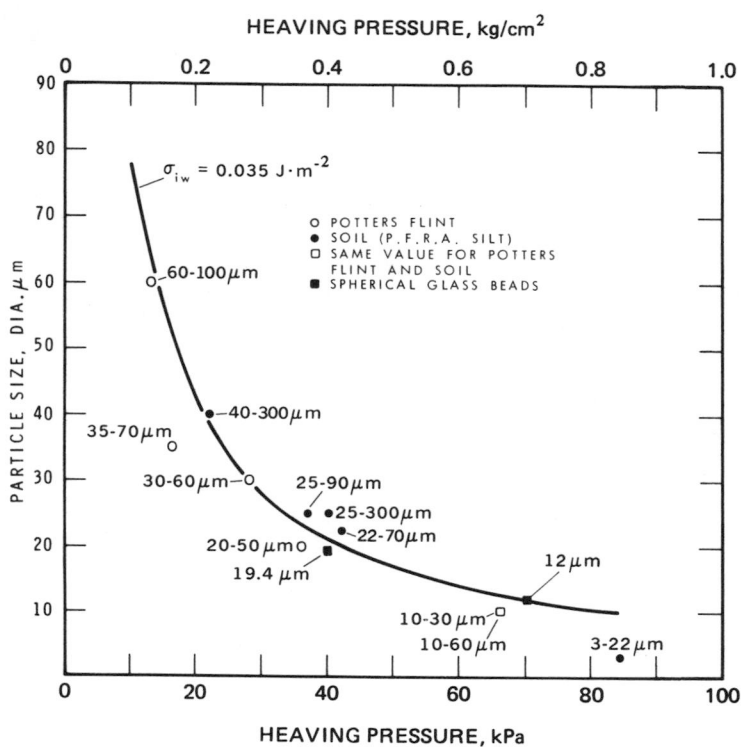

Fig. 5-2. Dependence of maximum pressure due to ice segregation on smallest particle size for various materials of given range in particle size (Penner, 1973) (Used with permission of the Organization for Economic Cooperation and Development.)

presents observations made in the laboratory on pressures developed during the freezing of particulate material of various grain size ranges (Penner, 1967, 1973). The specimens were confined and the frost action process allowed to proceed until the heaving pressure was maximum. Experiments such as these have demonstrated that the maximum force increases with decrease in the average grain size of the soil.

Figure 5-3 shows the effect on maximum pressure of increasing the

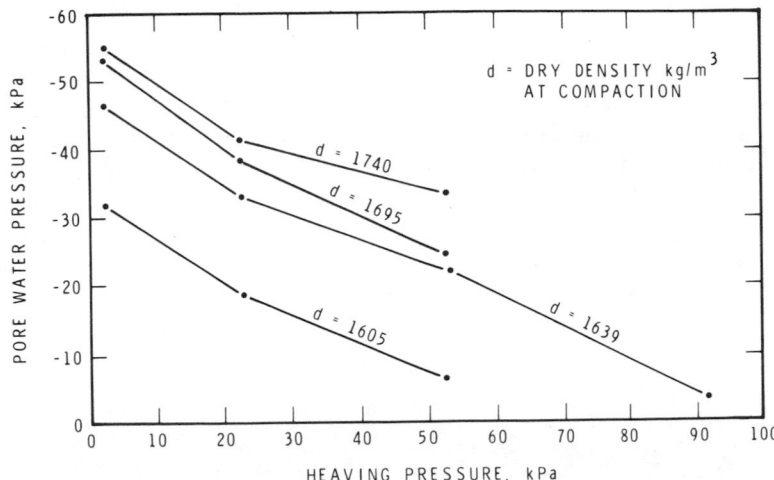

Fig. 5-3. Dependence of maximum pressure due to ice segregation on pore water pressure. From E. Penner (1959). The Mechanism of Frost Heaving in Soils, in Highway Research Board (Washington, D.C.) Bulletin *225*: 4. (Used with permission of Transportation Research Board.)

tension in water in the soil (Penner, 1959). It may be seen that the maximum pressure decreases almost linearly with increasing pore water tension.

One of the interesting features of water in porous materials is that not all of it freezes at the freezing temperature of bulk water. The amount of water that still remains unfrozen at a given temperature below freezing depends upon the material type and, in particular, upon its porosity (Williams, 1964; Litvan, 1978). In general, the amount of unfrozen water in a soil at a given temperature increases with decreasing grain size. For sand, significant amounts of unfrozen water are usually confined to temperatures above -1°C; for clay, there may still be significant unfrozen water a temperatures below -5°C. On thawing, the unfrozen water content curve is shifted to higher temperatures, indicating that on freezing the thermodynamic properties shift towards the values for bulk ice.

There has been growing evidence that the front of an actively growing ice lens does not coincide with the plane in which ice begins to form, but is on the cold side of it (Miller, 1973, 1978). Both fronts, called the ice front and the freezing front, respectively, are at temperatures less than 0°C, the degree depending on soil type, overburden pressure, pore water tension, and rate of freezing. An X-ray technique developed by Penner and Goodrich (1980) allows us to determine the position of ice while it is forming in soil in laboratory experiments. By spacing small temperature measuring devices along the length of the specimens, they were able to determine the position of the ice relative to the 0°C isotherm. Figure 5-4 is an example of what can be seen by this method. The 0°C isotherm is indicated in the figure and the position of two ice lenses clearly defined at A and B.

Using this technique it has been demonstrated that changing pressure causes the ice front to move. If a lens has been established and the pressure is subsequently increased, a new lens is initiated on the cold side of the old one and the old one deteriorates. Conversely, decreasing the pressure after a lens has been initiated causes a new one to be established on the warm side of the old one. In figure 5-4 a lens was formed initially at A. The pressure was subsequently reduced and a new lens initiated at B.

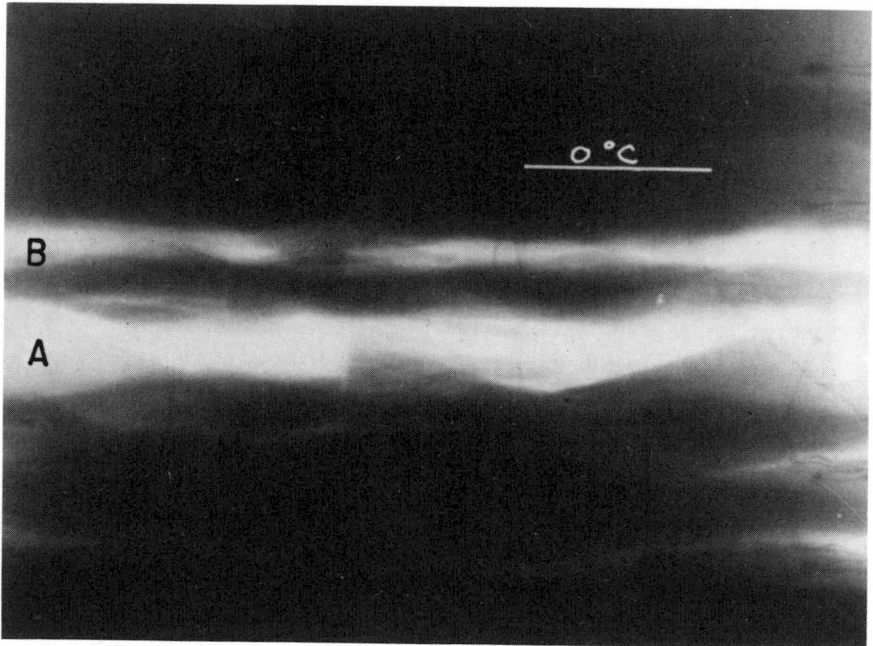

Fig. 5-4. X-ray photograph of ice lenses on cold side of 0°C isotherm (see Penner and Goodrich, 1980). See text for further explanation.

In strong porous materials such as rock and concrete, lens formation and associated heaving may not occur during ice segregation. Figure 5-5 presents a differential thermal analysis scan and associated length changes for small specimens of hardened cement paste. The freezing process began at about -8°C. As freezing progressed the specimen expanded, apparently

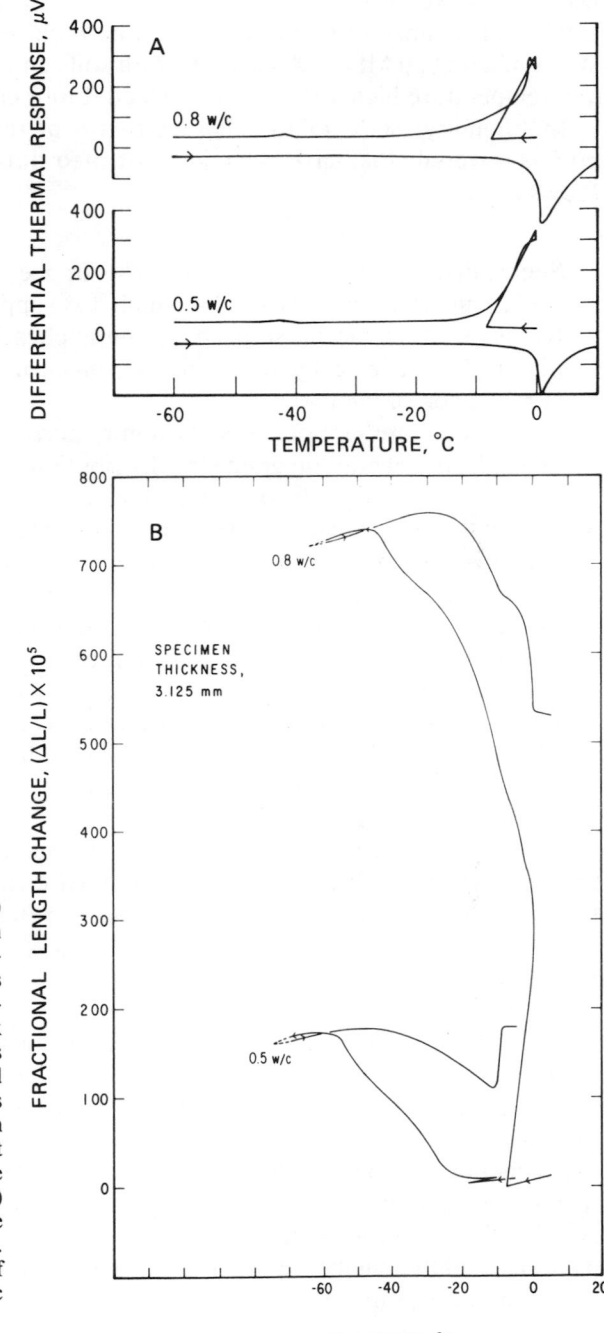

Fig. 5-5. Thermogram (A) and fractional length change (B) of vacuum saturated cement specimens determined simultaneously. The 3.2 mm thick x 32 mm diameter specimens were cooled to -60°C and rewarmed: w/c indicates the water/cement ratio in each sample. The apparent thawing at temperature above 0°C indicated in (A) is due to a temperature sensor (Litvan, 1972). (Used with permission of the American Ceramic Society.)

because the water moved out to the exterior surface. The expansion was small, in the range of strain that would be imposed on the material by a stress of about 50 MPa (7000 psi). As with soils on warming, the ice melted at a temperature higher than that at which it formed.

In summary, some of the characteristics of the freezing of water in porous materials that have been demonstrated through experiment are as follows:

1. Segregated ice may form as lenses. This implies a mobile film between the ice and the solid surfaces on which it is supported.
2. Ice segregation can be stopped by the application of pressure in the direction of freezing, by increasing the tension in the pore water, or by a combination of the two.
3. The applied pressure or water tension required to stop ice segregation depends, in general, on grain size distribution; the smaller the average grain size, the larger the pressure or tension.
4. Not all the water freezes at 0°C; the dependence of the unfrozen water content on temperature is determined by porosity, pore size distribution and nature of the material.
5. The freezing process occurs over a zone, the depth of which depends on the properties of the material, applied pressure, pore water tension and temperature gradient.
6. The location of the front of a growing ice lens may not coincide with the plane in which the ice first begins to form in the pores. Neither the ice front nor the freezing front need be at 0°C.

A model of the freezing process in porous materials must account for these observations. The six characteristics were clearly stated or implied in the combined contributions of Tabor (1930) and Beskow (1935).

THERMODYNAMIC MODEL

When Beskow stated in 1935 that a water film must exist between ice and soil particles for frost heaving to occur, a knowledgeable thermodynamicist probably could have stated the essence of the theoretical basis of the explanation of ice segregation that is now accepted. Fortunately for the experimentalist, the development and validation of a model usually requires a significant body of knowledge based on observation. This is certainly the case for ice segregation because of the variability of porous materials and the number of other independent factors that affect it. From the theoretical point of view, the greatest attention has been given to

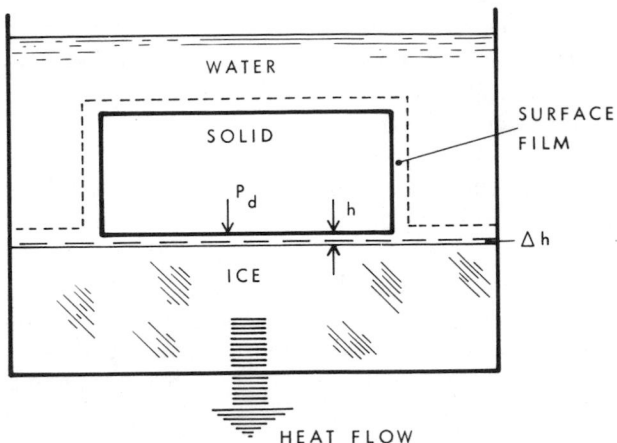

Fig. 5-6. Lifting of a solid with a plane ice-solid interface due to upward freezing of water.

describing the equilibrium condition for the water-porous solid system in terms of classical, steady-state thermodynamics.

Figure 5-6 is a simple representation of a flat solid resting on an ice surface in water that is freezing from the bottom upward. The ice is separated from the surface by a water film of thickness h. The thickness of this film is considered to be about 10^{-7} to 10^{-9}m (Hoekstra and Miller, 1967; Gilpin, 1978). A transition film also exists between the ice and water and the solid surface and water.

In principle, the thermodynamic properties of the ice, film and water can be determined. It is possible, therefore, to discuss the conditions governing equilibrium between them in terms of these properties. If the ice, film and water contain no impurities, a knoweldge of the temperature and pressure is sufficient to define thermodynamic equilibrium.

Two phases are in equilibrium when their chemical potentials are equal (Guggenheim, 1950, p. 31). If they are not equal, there will be a transfer of material from the phase of higher chemical potential to the phase of lower chemical potential. For the single chemical species system under consideration, it is usual to use the Gibbs-Duhem relation for the dependence of change in chemical potential on change in pressure and temperature, i.e.,

$$d\mu = VdP - SdT \qquad (1)$$

where V is the molar volume of the phase, S is the entropy per mole of phase, $d\mu$ is the change in chemical potential for a change of pressure, dP, and of temperature, dT.

The phenomena of interest occur over a relatively narrow temperature range near 0°C, and it is reasonable to assume for discussion purposes that V and S are constants.

Integrating equation (1) gives

$$\mu = \mu_o + V\Delta P - S\Delta T \tag{2}$$

where μ_o is the chemical potential for the pure phase at a reference state (e.g., 0°C and 1 atm pressure), and μ = chemical potential at temperatures and pressures that differ from their values at the reference state by ΔT and ΔP, respectively. The definition of equilibrium requires that μ_o for bulk ice be identical to that for bulk water, film and saturated water vapour. The temperature, pressure and other thermodynamic variables are relatively easy to measure for the bulk ice, water and vapour phases. This is not the case, however, for the film phase and often its properties must be implied indirectly.

It is useful to consider some of the observations on the behaviour of the film phase in order to develop an appreciation of its nature. It has been found for wettable materials that heat is evolved during the adsorption of an amount of water that would form a film about 7 to 10 molecules thick over all the surfaces; that is, a reaction occurs that binds the water to the surface.

A common characterization test for porous materials is to determine how their water content at a constant temperature depends on the pressure of the vapour with which they are in equilibrium. The change in chemical potential of water vapour, $\Delta\mu_v$, due to change in vapour pressure from the saturation value, P_o, to a lesser value, P, at constant temperature is given by

$$\Delta\mu_v = RT \ln(P/P_o) \tag{3}$$

where R is the gas constant and T is the temperature in °K. As P is less than P_o, the right hand side of equation (3) is negative, indicating a decrease in chemical potential.

Decreasing the vapour pressure also results in a reduction of water content. This correlation shows that under conditions of constant temperature the chemical potential of the adsorbed water film decreases with decreasing thickness. For equilibrium between the film and vapour phases after a change in vapour pressure at constant temperature, equation (2) requires that there be a change in the film pressure equal to

$$P_1 = (RT/V_1) \ln (P/P_o) \tag{4}$$

where V_1 = molar volume of the film fluid. Equation (4) implies that the decrease in film thickness is accompanied by a decrease in the fluid pressure in the film.

The film is the transition zone between the state of molecular order that exists at the solid-film interface and at the film-air interface, or in adjacent bulk water. The attraction between water molecules and the solid surface causes a pressure to exist in the film normal to the film-surface interface. This intrinsic pressure is a body force similar to the force due to gravity, but acting over a much smaller distance (Gilpin, 1978). It has been termed the spreading pressure, and like the depth dependence of the pressure in fluids due to gravity its value at the film-solid interface increases with increasing thickness of film.

Spreading pressure is not a common concept in geotechnical practice. By implication it is assigned the value 0 when the film is in the reference state, i.e., when the chemical potential μ in equation (2) equals μ_o. This can be appreciated from a consideration of the following common test procedure. The water content of a soil specimen can be controlled by placing the specimen on a porous plate connected to a water reservoir. When the pressure of the water in the reservoir is such that the pore water pressure at the soil-porous plate interface is atmospheric, the specimen will be saturated, neglecting the effect of gravity, and its pore water in equilibrium with vapour at pressure P_o. The water content can be reduced to the value in equilibrium with vapour at pressure P by reducing the pressure of the water in the reservoir by the amount P_1 in equation (4). As P_1 is negative and water is withdrawn from the specimen, the process is interpreted as applying a suction. What has actually occurred within the specimen, however, is that the film pressure at the solid-water interface has been reduced by P_1.

If two surfaces with adsorbed water films are brought together, it is found that the films are not easily displaced and that a repulsive force is mobilized. This force has been called the disjoining pressure (Derjaguin and Melnikova, 1958). The disjoining pressure P_d is defined as the negative of the pressure given by equation (4) (Padday, 1970). It is the pressure that would have to be applied to a film in equilibrium with vapour at the saturated pressure P_o to reduce its thickness to that which would be associated with vapour at pressure P. It is another manifestation of the forces responsible for the spreading pressure. Sources of these forces are considered to be van der Waals' attraction, the electrical double layer and orientation effects.

A similar transition region must exist on the surface of ice, and evidence has been found for it at temperatures as low as -10°C (Fletcher, 1972). The very fact that ice lenses can grow in fine grained soils when subjected to pressures as high as 10^6 N m^{-2} shows that the films can mobilize very significant disjoining pressures and still remain sufficiently fluid to sustain the growth process. Clearly, the film phase must have the ability to transfer a compressive stress between ice and a solid surface while transporting water in the freezing zone.

Theoretical and experimental studies have shown that the thickness of the film between ice and other solids depends on temperature, pressure between the ice and solid, and the pressure in the water (i.e., pore water pressure). As the film is in order of 10^{-8} m thick, it is clearly a most difficult task to establish the temperature and pressure dependence of the thickness other than in an indirect way.

Assume that heat flow from the ice-solid interface in figure 5-6 increases the thickness of the ice by Δh and, therefore, decreases the thickness of the film by the same amount. If the process occurs at constant temperature, there will be a corresponding decrease in the spreading pressure and in the chemical potential of the film. If the film is connected to adjacent films that have suffered no change, there will exist a difference in chemical potential and spreading pressures causing water to flow to the film and increase its thickness to the original value, displacing the solid upward.

If the solid is constrained from moving relative to the ice, the pressure must increase in the film. This results in an increase in its chemical potential toward the higher value of adjacent regions.

One of the controlling factors in the process is the resistance to the flow of water to the ice front. This resistance depends on the thickness of the film, the viscosity of water, the nature of the surfaces and the distance over which the water must flow. If the flow of water becomes a limiting condition, flow of heat away from the interface cannot be compensated by latent heat and the temperature of the interface must frop, thereby increasing the chemical potential and advancing the freezing zone.

Mackay and Burrous (1979) demonstrated that the ice segregation process can occur at relatively large flat surfaces. In one experiment they displaced a penny and a 2 cm³ polished granite cube upwards by an advancing ice front for a distance of some 2 to 3 cm, and stated that "many other solid objects were readily uplifted, particularly when the freezing rate was slow and the objects were raised on small supports in order to permit a thin layer of ice to form on the bottom before uplift commenced." They pointed out that this mode of frost action is of some importance during freeze-back of the active layer from below. Connell and Tombs (1971) measured directly a disjoining pressure of 20 kN m⁻² between an ice surface with the relatively large radius of curvature of 3 mm and a flat glass plate.

The shorter the distance water has to flow along the ice front the more readily can ice segregation be maintained. Freezing in porous solids offers such possibilities to reduce this flow distance. The porous structure also allows the mobilization of another important mechanism.

EFFECT OF CURVED INTERFACES

One of the characteristics of porous solids is that if the ice phase is to

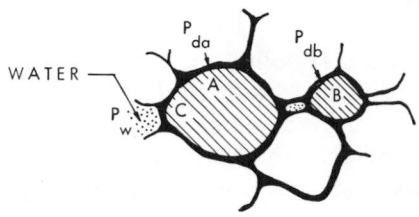

Fig. 5-7. Schematic representation of ice in pores; the heavy black lines represent the surface films. The effective radius of curvature of the ice at A = r_a, at B = r_b and at C = r_c.

grow it must propagate into and through pore spaces. Consider the ice-filled pores depicted in figure 5-7. Let the pore have an average radius of $\bar{r}_a$ at A and $\bar{r}_c$ at the opening C. Surface tension forces in the film on the curved surfaces cause pressure in adjacent ice. The dependence of this pressure, P_r, on the average radius of curvature, $\bar{r}_a$, of the ice surface at A is given by (Everett, 1961)

$$P_r = 2e_{iw}/\bar{r}_a \tag{5}$$

where e_{iw} is the surface energy of the ice-water interface, and P_r is positive for a concave ice surface. If the disjoining pressure between the ice and the adjacent solid surface is P_{da}, the total change in pressure from the reference state in the ice at the curved interface is

$$\Delta P_i = P_{da} + 2e_{iw}/\bar{r}_a \tag{6}$$

Substituting this value for ΔP in equation (2) gives for the chemical potential of the ice at the interface at A

$$\mu_{ia} = \mu_o + V_i(P_{da} + 2e_{iw}/\bar{r}_a) - S_i\Delta T_{ia} \tag{7}$$

If the pore water pressure at C is P_w, the pressure on the ice at C is

$$P_i = P_w + (2e_{iw}/\bar{r}_c) \tag{8}$$

The chemical potential of the water at C is

$$\mu_w = \mu_o + V_w P_w - S_w\Delta T_w \tag{9}$$

and of the ice

$$\mu_{ic} = \mu_o + V_i(P_w \, 2e_{iw}/\bar{r}_c) - S_i\Delta T_{ic} \tag{10}$$

For equilibrium $\mu_w = \mu_{ia} = \mu_{ic}$. Equating equations (7) and (10) and assuming $\Delta T_w = \Delta T_{ia} = \Delta T_{ic} = \Delta T$ gives

$$P_{da} + (2e_{iw}/\bar{r}_a) = P_w + (2e_{iw}/\bar{r}_c) \tag{11}$$

Equating equations (9) and (10) gives

$$V_i(P_w + 2e_{iw}/r_c) - V_w P_w = (S_i - S_w)\Delta T \tag{12}$$

For T small (e.g., less than -5°C) it is usually assumed that

$$S_i - S_w = -L/T_o \tag{13}$$

where L = latent heat of fusion for water; T_o = the melting point of bulk ice = 273.18 °K.

Therefore, at equilibrium

$$V_i(P_w + 2e_{iw}/\bar{r}_c) - V_wP_w = -L\Delta T/T_o \tag{14}$$

Equations (8), (11) and (14) define the conditions for which ice will propagate out of a pore of effective radius $\bar{r}_a$ through effective radius $\bar{r}_c$ (i.e., ice will propagate through all pore openings of average radius greater than or equal to $\bar{r}_c$). They indicate that a reciprocal relation should exist between the applied load and the pore water pressure, as observed in laboratory experiments.

Koopmans and Miller (1966) utilized equations (5) and (14) in a comparison of the pressure dependence of the water content of a saturated soil free of colloidal material (determined in a pressure plate apparatus) and the temperature dependence of the unfrozen water content in the same soil. Williams (1966) used equation (8) as the basis for a comparison of the air pressure required to initiate air intrusion into an unfrozen saturated soil and the conditions for the penetration of ice into pores at the freezing front for the same soil. The theoretical relation that applies when the mean curvatures of the respective air-water or ice-water interfaces are the same, i.e.,

$$(P_a - P_w)/(P_i - P_w) = e_{aw}/e_{iw} \tag{15}$$

was confirmed by experiment. In equation (15) P_a = air pressure applied in the pressure plate test or in the air entry test, P_i = pressure exerted by the ice, P_w = pore water pressure at the ice- or air-water interface, e_{aw} is the surface energy of the water-air interface. These investigations, particularly that of Koopmans and Miller (1966), demonstrate the equivalent effect of pore air pressure and pore ice on the water films with which they are in contact.

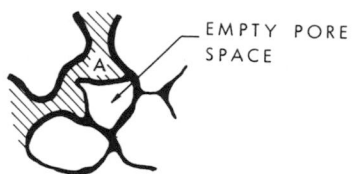

Fig. 5-8. Schematic representation of ice growing into a pore; the heavy black lines represent the surface films. If the temperature is sufficiently low, the pore will spontaneously fill with ice, even if it is not initially filled with water.

Consider figure 5-8, which portrays ice advancing into a pore. Miller (1973), and Bresler and Miller (1975) presented the conditions for which ice will exist in reentrants and similar cavities associated with pores. As the ice phase grows, a metastable condition is attained at which the pore will spontaneously fill with ice, whether it was previously filled with water (saturated) or not (unsaturated). This implies that a state is reached for which the chemical potential of ice in a filled pore is less than that in a partially filled pore. Such a condition occurs when ice advancing from adjacent crevices coalesces or when the ice front from a single source has advanced to a particular position. In both cases the ice surface in the pore reaches a position at which, with further growth, its curvature decreases and may even reverse, as illustrated at A in figure 5-8, resulting in a decrease in P_r in equation (5) and a lowering of the chemical potential.

This possibility of spontaneous filling of space by ice has very significant implications for porous materials. As the zero degree isotherm surface advances through the material, the position of the freezing front will be at the temperature at which ice will propagate through the maximum size openings (pores or cracks). Behind this front pores will spontaneously fill with ice when their temperature reaches a value determined by the pore geometry and pore water pressure. It can be expected that if water is available to a porous material on the cold side of the 0°C isotherm, the material will become saturated with ice (Bresler and Miller, 1975). This agrees with field and laboratory observations of the ice-saturated state of soils in which water can flow to the freezing zone during the freezing process.

INITIATION OF ICE LENSES

A question of some interest is: what are the conditions necessary to initiate an ice lens? Consider again the schematic representation of pores in figure 5-7. Assume that in the thawed state the soil is subjected to an effective stress of $P_e = P - P_w$ where P is the total stress on a horizontal plane. For the purpose of this discussion, assume that ice is present in pores A and B having effective radii $\bar{r}_a$ and $\bar{r}_b$, respectively. Let $\bar{r}_a$ be larger than $\bar{r}_b$. If the pores are interconnected and in thermodynamic equilibrium at the same temperature, then from equation (7)

$$P_{da} + (2e_{iw}/\bar{r}_a) = P_{db} + (2e_{iw}/\bar{r}_b) \tag{16}$$

where P_{da} and P_{db} are the disjoining pressures imposed on the ice through the film by the adjacent solid surfaces. P_{da} must be larger than P_{db}.

Consider now the film phase. The total pressure in the film, P_t, is the sum of the local intrinsic or spreading pressure, P_s, and the imposed or disjoining pressure, P_d, i.e.,

$$P_t = P_s + P_d \tag{17}$$

Since the film at A is in equilibrium with the films at B and C (no flow), $P_{ta} = P_{tb} = P_{tc}$. For this to occur, any change in disjoining pressure must be compensated by an equal but opposite change in spreading pressure. It should be re-emphasized that the spreading pressure is felt only within the film.

The change from the reference state in the film pressure at C, where the disjoining pressure is zero, is equal to the pore water pressure P_w. For the equilibrium condition, therefore, the change in the total film pressure from the standard state at A and B is also P_w and the equivalent of equation (12) is

$$V_i(P_d + 2e_{iw}/\bar{r}_c) - V_1P_w = (S_i - S_1)\Delta T \tag{18}$$

If it is assumed that $(S_i - S_1) = -L_1/T$, where L_1 is the latent heat released in the freezing of film water at temperature T°K, the equivalent of equation (14) is

$$V_i(P_d + 2e_{iw}/\bar{r}_c) - V_1P_w = -L_1\Delta T/T \tag{19}$$

Equation (19) shows that for a surface of constant temperature the disjoining pressure that exists between the ice and adjacent solid surface for each pore is determined by the temperature, the effective radius of the pore, and the pore water pressures. At equilibrium, where there is no flow, the total pressure in the film must equal that of the films in the unfrozen zone at the freezing front corrected for the effect of gravity.

When the average of the disjoining pressures over a constant temperature surface equals the effective stress, P_e, the condition is established for separation and the initiation of a lens. Equation (19) shows that it depends on the pore water pressure and the applied or overburden pressure through their effect on the required value for P_d

This model for ice segregation has some interesting implications. Equation (19) indicates that the smaller the pore size the smaller is the disjoining pressure required for equilibrium and, therefore, the thicker is the film. This, along with the increasing surface area associated with decreasing pore size, should increase the possibilities for water to move through the freezing zone. Once a lens has been initiated and as long as permeability is not a limiting factor, it is found that the finer the pore system the more readily can water move to regions of the ice front where the local effective radius of curvature is large and, therefore, the disjoining pressure is large. A negative radius of curvature such as that over a soil particle would require an even larger disjoining pressure for equilibrium.

An extreme situation would be a flat surface with many fine holes over which essentially the full disjoining pressure of which the water film is capable can be mobilized. Such a situation could be approached in clays for which the plate-like particles have a preferred orientation. Assuming $\Delta T = -2°C$, $L_1 = 335 \times 10^{-3}$ J kg^{-1}, $\bar{r} = \infty$, $P_w = 0$ (saturated soil), T = 270°K, and $V_i = 1.1 \times 10^{-3}$ m³ kg^{-1}, equation (19) gives $P_d = 2.26 \times 10^6$ N m^{-2} (330 psi). This is in the same range as the maximum heaving pressures observed for clay soils.

Increasing the pore water tension is a more severe limiting condition for ice segregation. In soils, the openings to the ice front vary in size. As the pore water tension increases, pores begin to empty of water and ice, beginning with the largest (Bresler and Miller, 1975). This restricts their ability to deliver water to the ice front, if it exists, and to contribute to the total disjoining pressure.

For a growing ice lens the total film pressure at the ice front is less than that in the unfrozen water at the freezing front, and this pressure difference along with the permeability of the freezing zone will determine the rate of flow to the ice lens. The lens will continue to grow until the total film pressure at the ice front becomes equal to that in the unfrozen water through either an increase in the disjoining pressure or a decrease in the pore water pressure, or until a new lens is initiated at another location.

Of particular interest are porous materials such as porous rock or concrete that are too strong to allow an ice lens to be initiated. If water-saturated material is subjected at one face to freezing conditions, the absorbed water will experience a force that moves it to regions where ice has formed (e.g., on the surface or in large pores). Assume that a temperature gradient exists such that not too far from the ice front the material is saturated at 0°C and that initially there is space for ice to form without imposing a force on surrounding walls. As freezing in such fine pore materials may not begin until the temperature is in the range of -10°C, it is easy to imagine a situation for which ΔT in equation (19) could be -5°C or lower. A difference in film pressure of the order of 10^7 Nm^{-2} could conceivably be introduced between the ice front and the saturated region at 0°C. As the ice would come into contact with the surrounding cavity walls, this difference in spreading pressures would gradually be reduced by the development of disjoining pressures.

If cavities are spaced sufficiently closely and are large enough to provide storage for the water present in the solid, damaging spreading pressures can be dissipated and disjoining pressures avoided. This is the basis for air entrainment of concrete and a recent invention by Litvan and Sereda (1978) that involves the mixing of coarse pored paticles with normal aggregate.

NON-EQUILIBRIUM EFFECTS

A most difficult question to answer is whether thermodynamic equilibrium is ever established in the field or in the laboratory in porous ice-filled materials subjected to a gradient of temperature and pressure. Certainly experiments can be run in the laboratory for which dimensional changes due to ice freezing or thawing are either zero or less than the resolution of the measuring method. Such experiments are run usually for a period of days at the most and cannot show, for example, whether frozen soil and rock formations subjected to natural or imposed conditions will thaw or heave significantly or experience redistribution of ice over a period of years or centuries. This question is of great interest with respect to the stability of the ground ice that has absorbed so much of Ross Mackay's attention, but it has not been studied widely (Harlan, 1974).

Consider the relatively simple case of a spherical particle embedded in ice (figure 5-9a). Assume that it and the ice are initially at a constant negative temperature T_a, that a gradient is imposed such that the temperature at a remains the same, and that at b on the opposite side becomes $T_b <$ T_a. The chemical potential of the film at b is now greater than that of the adjacent ice, and some of the water is converted to ice. If the temperature change is maintained, according to equation (19) the pressure in the film at b and the adjacent ice must increase to re-establish equilibrium. This pressure is felt at a, resulting in the melting there of some of the ice. The net effect is a transfer of water in the cold direction and of the particle in the warm direction. Experiments by Hoekstra and Miller (1967) and by Romkens and Miller (1973) have demonstrated that this indeed occurs. The rate of movement is affected by several variables, amongst them the temperature gradient, the nature, shape and size of the particle, and its average temperature. Significant movement occurs only for small particles and average temperature greater than -2°C.

A better known phenomenon is regelation (Nye, 1967; Drake and Shreve, 1973). Consider a wire drawn through ice at constant temperature as shown in figure 5-9b. Because of the pressure between the ice and the

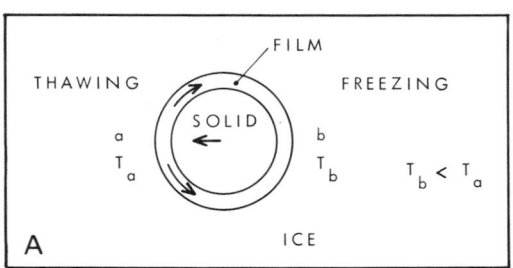
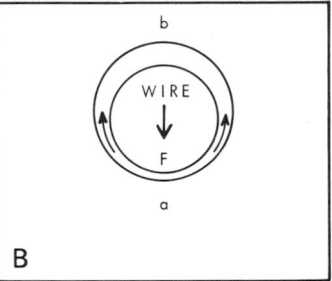

Fig. 5-9. (a) Displacement of a solid particle by temperature induced flow of water in the film from the warm to the cold side.
(b) Movement of a wire through ice by regelation due to a force, F. Ice melts at a, flows in the film to b and refreezes.

wire the chemical potential of the ice at a is greater than that of the adjacent water film, and melting occurs. The wire moves downward, reducing the pressure at b, and the pressure gradient that has now been imposed on the film forces water to the back of the wire. As the film at b now has a larger chemical potential than the adjacent ice, water from it freezes. Factors that control this effect are the type and size of wire, the applied pressure, the average temperature of the wire, the viscosity of water and the resistance to flow around the wire. Significant rates of regelation under laboratory conditions are also confined to temperatures close to the melting point. Although the relative rate of movement of ice and particles due to temperature gradients and regelation may be small, the total movement may be significant over geological periods of time.

Pressure gradients may also result in thermodynamic instability, tending to move ice from positions of high stress to positions of low stress, resulting in significant redistribution. Although this effect is of potential importance for many field situations, it has received little attention.

CONCLUSIONS

In this presentation I have tried to give an appreciation of the interesting behaviour of ice and water in porous materials. This behaviour is confined primarily to the temperature range of 0 to -1°C for coarse grained soils such as sand; 0 to -5°C for fine grained soils such as clay; and perhaps to temperatures as low as -10°C for rocks and similar fine pore materials. Below -10°C ice in porous materials is more inert, more rock-like in its behaviour, and from Ross Mackay's point of view probably less interesting except with respect to distribution and quantity.

The thickness of frozen ground in the temperature ranges of 0 to -1°C and 0 to -5°C depends on the temperature gradient. For deposits subjected to the geothermal gradient of 3×10^{-2}°C m^{-1} it can extend to over 150 m. These are relatively extensive thicknesses of thermodynamically sensitive deposits.

A major challenge now is to establish the dependence of the rate of segregation and of the redistribution of ice on the factors that determine them. Another interesting question is: if the combined overburden and pore water pressures are such as to not totally suppress ice segregation, can there ever be true thermodynamic equilibrium in the extensive thickness of ice-saturated soils that exists naturally? If not, what is the significance of this on a geologic time scale, particularly for deposits subjected to long-period temperature and stress changes? These are questions of some geologic and engineering interest and ones that Ross Mackay may wish to tackle.

REFERENCES

Beskow, G. 1935. Soil freezing and frost heaving with special application to roads and railways. Stockholm, Norstedt and Soner: 242 p. (Translation by J.O. Osterberg, Technological Institute, Northwestern University, 1947).

Bresler, E. and Miller, R.D. 1975. Estimation of pore blockage induced by freezing of unsaturated soil. Amer. Geophys. Union, Task Force of the Div. of Hydrology Conf. on Soil-Water Problems in Cold Regions, Calgary. Proc. (mimeo.): 161-175.

Connell, D.C. and Tombs, J.M.C. 1971. The crystallization pressure of ice — a simple experiment. J. Glaciology *10*: 312-315.

Derjagiun, B.V. and Melnikova, N.K. 1958. Mechanism of moisture equilibrium and migration in soils. Highway Res. Bd., Washington. NAS/NRC (USA). Spec. Rep. *40*: 43-64.

Drake, L.D. and Shreve, R.L. 1973. Pressure melting and regelation of ice by round wires. Roy. Soc. London, Proc. *A332*: 51-83.

Everett, D.H. 1961. The thermodynamics of frost damage to porous solids. Faraday Soc., Trans. *57* (465, Part 9): 1541-1551.

Fletcher, N.H. 1973. The surface of ice. *In* Whalley, E., Jones, S.J. and Gold, L.W., eds., Physics and Chemistry of Ice, Papers from a Symposium, Ottawa, 14-18 August, 1972. Roy. Soc. Canada.: 132-136.

Gilpin, R.R. 1978. A model of the liquid-like layer between ice and a substrate with applications to wire regelation and particle migration. J. Colloid and Interface Science *69*: 235-251.

Guggenheim, E.A. 1950. Thermodynamics. 2nd ed. Amsterdam, North-Holland and N.Y., Interscience Publishers: 412 pp.

Harlan, R.L. 1974. Dynamics of water movement in permafrost — a review. *In* Permafrost hydrology. Workshop/Seminar, Calgary, Canada, Feb. 18-20. Can. Nat. Committee, Int. Hydrol. Decade: 69-77.

Hoekstra, P. and Miller, R.D. 1967. On the mobility of water molecules in the transition layer between ice and a solid surface. J. Colloid and Interface Science *25*: 166-173.

Koopmans, R.W.R. and Miller, R.D. 1966. Soil freezing and soil water characteristic curves. Soil Science Soc. Amer., Proc. *30*: 680-685.

Litvan, G.G. 1972. Phase transitions of adsorbates IV, mechanism of frost action in hardened cement paste. J. Amer. Ceram. Soc. *55*: 38-42.

Litvan, G.G. 1978. Adsorption systems at temperature below the freezing point of the adsorptive. Adv. Colloid and Interface Science *9*: 253-302.

Litvan, G.G. and Sereda, P.J. 1978. Particulate admixture for enhanced freeze-thaw resistance of concrete. Cem. Concr. Research *8*: 53-60.

Mackay, J.R. 1971. The origin of massive ice beds in permafrost, western Arctic coast, Canada, Can. J. Earth Sciences *8*: 397-422.

Mackay, J.R. 1972. The world of underground ice. Assoc. Amer. Geographers, Ann., *62*: 1-22.

Mackay, J.R. 1973. The growth of pingos, western Arctic coast, Canada. Can. J. Earth Sciences *10*: 979-1004.

Mackay, J.R. 1974. Ice wedge cracks, Garry Island, N.W.T. Can. J. Earth Sciences, *11*: 1366-1383.

Mackay, J.R. 1979. Pingos of the Tuktoyaktuk Peninsula Area, Northwest Territories. Géographie Phys. Quat. *33*: 3-61.

Mackay, J.R. and Burrous, C. 1979. Uplift of objects by an upfreezing ice surface. Can. Geotech. J. *16*: 609-613.

Miller, R.D. 1973. Soil freezing in relation to pore water pressure and temperature. Permafrost 2nd Int. Conf., Yakutsk, U.S.S.R., July 13-28: North American Contribution. Washington, D.C. NAS (U.S.A.): 344-352.

Miller, R.D. 1978. Frost heaving in non-colloidal soils, 3rd Int. Conf. on Permafrost, Edmonton, Canada, July 10-13. Proc. *1*. Ottawa, NRC Canada Pub. *16529*: 707-713.

Nye, J.F. 1967. Theory of regelation. Phil. Mag. *16*: 1249-1266.

Padday, J.F. 1970. Cohesive properties of thin films of liquids adhering to a solid surface. Faraday Soc., Spec. Disc. *1*: 64-74.

Penner, E. 1959. The mechanism of frost heaving in soils. Highway Res. Bd., Washington. Bull. *225*. NAS/NRC (USA), Pub. *685*: 1-22.

Penner, E. 1967. Heaving pressure in soils during unidirectional freezing. Can. Geotech. J. *4*: 398-408.

Penner, E. 1970. Frost heaving forces in Leda Clay. Can. Geotech. J. *7*: 8-16.

Penner, E. 1973. Frost heaving pressures in particulate materials. Paris, OECD, Symp. on Frost Action on Roads, *1*: 379-385.

Penner, E. and Goodrich, L.E. 1980. Location of segregated ice in frost susceptible soil. Presented at 2nd Int. Symp. on Ground Freezing, Trondheim, Norway. June 24-26. Preprints: 626-639.

Romkens, M.J.M. and Miller, R.D. 1973. Migration of mineral particles in ice with a temperature gradient. J. Colloid and Interface Science *42*: 103-111.

Tabor, S. 1930. The mechanics of frost heaving. J. Geology *28*: 303-317.

Williams, P.J. 1964. Unfrozen water content of frozen soils and soil moisture suction. Geotechnique *14*: 231-246.

Williams, P.J. 1966. Pore pressures at a penetrating frost line and their prediction. Geotechnique *16*: 187-208.

6
MODELS OF SOIL FREEZING

M.W. Smith

Geotechnical Science Laboratories, Department of Geography,
Carleton University, Ottawa, Ontario, K1S 5B6 Canada

ABSTRACT

Although permafrost is, by definition, a thermal condition, the properties of and processes in frozen ground are substantially complicated by the presence of water. This paper reviews models that attempt to explain the movement of water and formation of ice lenses in freezing soils. The capillary model appeals to the capillary suction present at an ice/water interface at the freezing plane to move water toward a growing ice lens. Experimental and field evidence shows, however, that capillary suction is insufficient to explain continued ice segregation under high overburden pressures. The hydrodynamic model recognises the interrelatedness of heat and moisture transfers. In this model changes in hydraulic conductivity in the freezing zone are seen to lead to segregated ice development behind the freezing front where a much greater difference in pressure states exists between ice and water. The secondary frost heave model further develops the analysis to include consideration of stress conditions and overburden pressure. Neither of the latter two models can be applied in the field in routine calculations because of information requirements and natural variability in soils. The paper closes, then, with some practical considerations. In this subject, Ross Mackay's observations of segregated ice development in the field are of critical importance.

RÉSUMÉ:
Modèles de congélation du sol

Quoique le pergélisol est, par définition, un état thermique, les propriétés du gélisol et les processus dans le gélisol sont compliqués substantiellement par la présence de l'eau. Cet article est un compte-rendu des modèles qui tentent d'expliquer le mouvement de l'eau et la formation des lentilles de glace dans les sols lors de l'engel. Le modèle capillaire fait appel à la tension capillaire à l'interface glace-eau au plan d'engel pour attirer l'eau vers une lentille de glace en croissance. Cependent, des données de terrain et les résultats expérimentaux indiquent que la tension capillaire ne peut expliquer une ségrégation continue de glace en présence de pressions élevées dans le sol. Le modèle hydrodynamique comprend les interrelations entre les transports de chaleur et d'humidité. Dans ce modèle, des changements de conductivité hydraulique dans la zone d'engel aboutissent au développement de glace de ségrégation en arrière du front d'engel où existe une plus grande différence de pression entre la glace et l'eau. Le modèle secondaire de soulèvement élabore cette analyse pour incorporer des considerations de tensions et de pression dans le sol. Ni l'un ni l'autre de ces deux modèles ne peut être utilisé de façon routinière pour des calculs sur le terrain à cause des données requises et des variations naturelles des sols. L'article termine sur des considérations pratiques. Dans ce domaine, les observations de terrain de Ross Mackay sur la développement de la glace de ségrégation sont d'une importance critique.

МОДЕЛИ ЗАМЕРЗАНИЯ ГРУНТА

М. В. СМИТ

РЕЗЮМЕ

Хотя, согласно определению, многолетняя мерзлота это тепловое явление, характеристики и процессы в замерзшем грунте существенно усложняются присутствием воды. В настоящем докладе рассматриваются модели пытающиеся выяснить движение воды и формирование ледяных линз в замерзающем грунте. Капиллярная модель видит в капиллярном всасывании, происходящем на границе лед/вода на замерзающем уровне, возможность продвижения воды к растущей ледяной линзе. Экспериментальные и полевые данные указывают, однако, что капиллярное всасывание не достаточно, чтобы выяснить беспрерывную сегрегацию льда при больших давлениях покрывающей породы. Гидродинамическая модель принимает во внимание взаимосвязь при передаче тепла и влаги. В этой модели, перемены гидравлической проводимости в замерзающей зоне приводят к развитию сегрегационного льда за промерзающим фронтом, где существует гораздо большая разница уровня давления между льдом и водой. Модель повторного мерзлотного пучения развивает анализ дальше, включая вопрос условий напряжения и давления производимых покрывающей породы. Последние две модели не применимы в полевых работах для повседневных расчетов из за недостатка информационных данных и естественной разнообразности грунта. Доклад заключается рассмотрением практических вопросов. Здесь наблюдения Росс Макая в области развития сегрегационных льдов в полевых условиях являются весьма важными.

INTRODUCTION

Inasmuch as permafrost is defined on the basis of temperature, the analysis of soil thermal regimes is of obvious importance to a variety of studies. However, whilst much can be gained from the simple application of heat conduction theory, the purely thermal approach fails to consider the real complexity of the soil freezing process. This complexity arises from the now well-known fact that "frozen" soils can contain appreciable amounts of unfrozen water at temperatures down to several degrees below 0°C. The amount depends upon temperature and soil type, and the relationship is

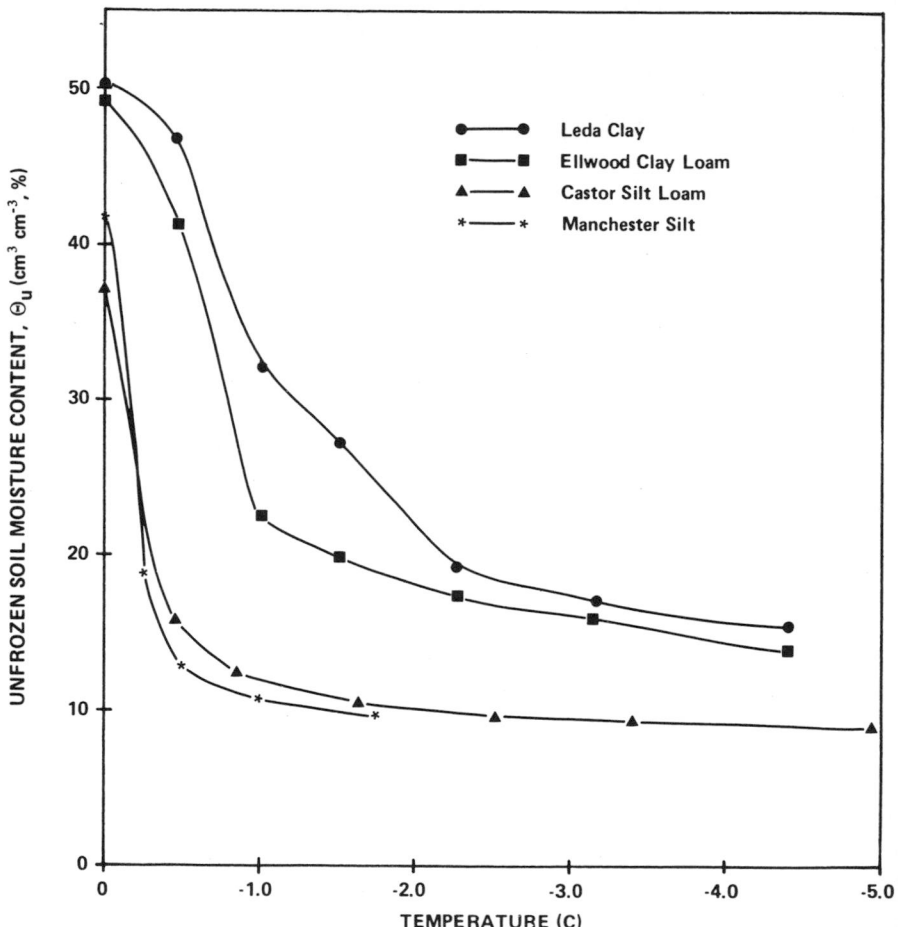

Fig. 6-1. Unfrozen water contents for various soils (from Patterson and Smith, 1981)

illustrated in figure 6-1. Changes in temperature will alter the phase composition of frozen soils and can cause dramatic changes in physical and mechanical properties. In addition, a temperature gradient in frozen soil is analogous to a water potential gradient in unfrozen soil, and thus temperature gradients imply water movements.

When a soil freezes, moisture may be redistributed due to potential gradients, and may subsequently freeze. Frost heave occurs when ice accumulation exceeds the volume of soil pores, resulting in an ice-rich zone, or discrete layers of ice called ice lenses. Such ice segregation and frost heave invariably accompany the freezing of finer-grained soils, provided that water is present. The vertical displacement of the ground and the generation of high heaving pressures are troublesome and complex problems faced in northern engineering. Such forces are also responsible for a variety of natural features in cold regions (Mackay, 1971, 1979).

The segregation of ice lenses during soil freezing is a complicated and challenging problem not yet fully understood. A comprehensive analysis must deal with the coupling of heat and moisture flows, and there are complicated relationships involved. Whilst it is true that certain soils are more frost susceptible than others, frost heave is not some intrinsic soil property, but rather the result of the combined thermal, hydrologic and stress conditions in the soil. Certain soil properties are obviously important, such as the relative amounts of ice and water in the pore space, and the relationship between thermal and hydraulic conductivities. These relationships vary for each soil and can only be determined by experiment.

This paper reviews various models that attempt to explain the movement of water and formation of ice lenses in freezing soils. They are commonly known as the capillary, hydrodynamic and secondary heave models. Reference is made to significant field and laboratory evidence, and the importance of frost heave to northern pipelines is also briefly discussed. The paper concludes with a brief discussion of a more practical approach to frost heave prediction.

One should appreciate that the structure and behaviour of frozen soil *in situ* can differ markedly from that of laboratory samples. Therefore, the application of results from laboratory experiments to problems in practice is not straightforward. Field studies result in an awareness of the limitations of laboratory and theoretical studies in practice, and in this regard the contribution of Ross Mackay is notable.

GROUND THERMAL ANALYSIS

As a freezing front descends in the soil, one of two things can happen. Pore water may freeze *in situ* as growing ice crystals invade and fill the soil pores. In this case the development of freezing pressure results from the 9 percent volume expansion of water upon freezing. Alternatively, segregated ice lenses may form in the vicinity of the freezing front as a result of soil moisture migration. In this case, the heaving pressure, which is derived from the difference in pressure states between the ice and water, can be several atmospheres.

Numerical models of soil freezing may be divided, correspondingly, into two groups:

i) Moisture content is assumed static and only the thermal aspect (conduction) is considered.
ii) The flows of heat and moisture are considered simultaneously, via a system of coupled equations.

Soil thermal analysis has conventionally followed the first approach, which may be acceptable in many instances. To review this briefly, the thermal regime is described by the thermal diffusion equation

$$\frac{\partial}{\partial z} K(T) \frac{\partial T}{\partial z} = \frac{\partial}{\partial t} (C(T) T) \qquad (1)$$

where K (thermal conductivity) and C (volumetric heat capacity) adopt frozen or unfrozen values as appropriate. In numerical models, the latent heat of freezing is usually managed by an "apparent heat capacity" term, which replaces C(T) in (1):

$$C_a(T) = C_m(T) + L\rho_w (d\Theta_u/dT)_T \qquad (2)$$

where C_m is the volumetric heat capacity of the soil-ice-water mixture, L is the latent heat of freezing (0.334 MJ kg^{-1}), ρ_w is the density of water, and $(d\Theta_u/dT)_T$ the slope of the freezing characteristic curve (figure 6-1) at temperature T. Freezing characteristic data are only now becoming routinely available (Patterson and Smith, 1981), and, previously, many numerical models made simplifying assumptions in equation (2). If K(T) were known (e.g. figure 6-2), this relationship could also be incorporated into equation (1). However, there is virtually no information available on the thermal properties of freezing soils in the range 0° to -3°C.

Equation (1) describes, in a general way, the flow of heat in soils; its application has been reviewed by Gold and Lachenbruch (1973), Harlan and Nixon (1978) and Goodrich and Gold (1981). The rest of this paper is concerned with soil freezing when water migration is involved.

THE CAPILLARY MODEL OF ICE SEGREGATION

The capillary model was the first and is the simplest model to account for the development of ice lenses in freezing soils. According to this theory it is

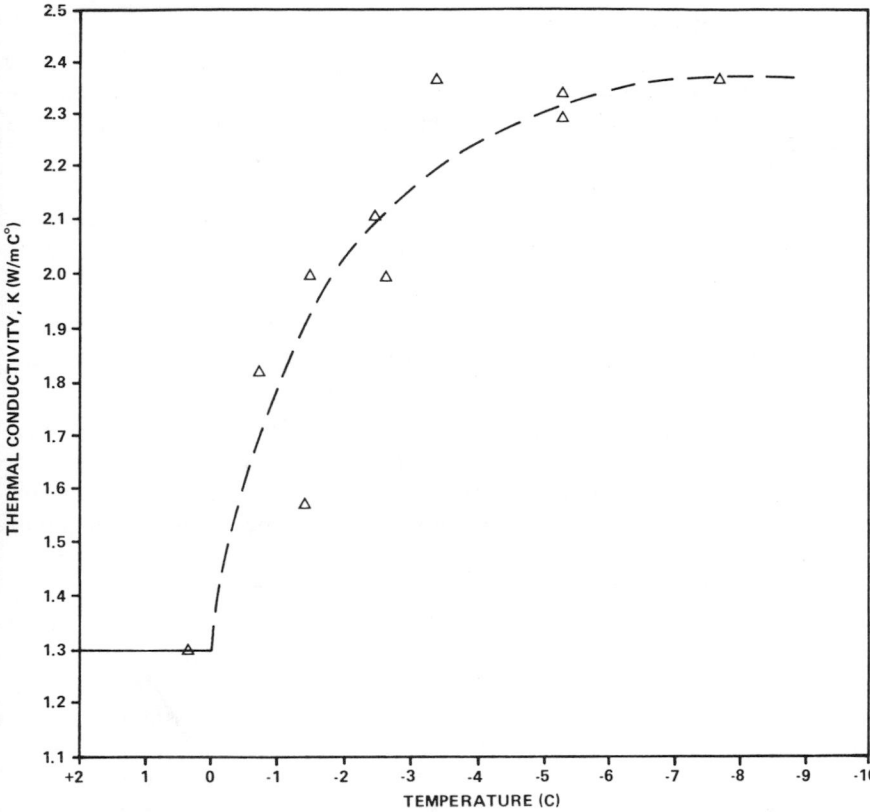

Fig. 6-2. Thermal conductivity of Castor silt loam at various freezing temperatures (from Riseborough *et al.*, 1983)

the capillary suction present at an ice/water interface that causes water to move towards a growing ice lens. The capillary model outlines the thermo-dynamic requirements for ice lens growth but does not function as a truly predictive model. Although certain aspects of the model remain correct, it does not constitute a comprehensive theory of soil freezing. A description of the theory can be found in Penner (1959) or Williams (1967).

Williams (1967) demonstrated that as the temperature falls in a freezing soil, the pressure on the water phase (P_w) will fall below atmospheric (i.e. a suction develops), while the ice pressure (P_i) is assumed to be atmospheric. In this case the clapeyron equation has the form

$$P_i - P_w = -(T - T_o)L/V_wT_o \qquad (3)$$

where T_o is the freezing temperature of pure free water (273.15° K), T is the temperature of the system and V_w is the specific volume of water (0.001 m³ kg⁻¹). Everett (1961) showed that small crystals in their own melt adopt a form that satisfies the equation

$$P_i - P_w = 2s_{iw}/r_{iw} \qquad (4)$$

where s_{iw} is the interfacial surface tension ($3.05 \times 10^{-2} N/m$) and r is the radius of the interfacial curvature. Combining (3) and (4) we can derive the following:

$$r_{iw} = 2s_{iw}V_wT_o/L(T - T_o) \qquad (5)$$

Thus, as the soil freezes and T decreases, r_{iw} becomes smaller, allowing ice to penetrate smaller and smaller pore spaces. At any instant we can visualize either of two things happening:

i) If r_{iw} is smaller than the minimum pore size in the soil (r_p), the ice invades and fills the soil pores and the freezing front advances into the soil.

ii) If, however, the freezing front encounters soil pores for which $r_{iw} > r_p$, the ice is unable to penetrate and the freezing front pauses.

In case (ii), since the water at the freezing front is then at a lower pressure than the water in the unfrozen soil below, a suction (potential) gradient will be established and water will flow towards the front. Accumulation of water there will cause a rise in P_w and thus some water must freeze, forming an ice lens. Continued growth of the lens depends upon the balance between the rate at which liberated latent heat can be removed (i.e. the thermal conditions in the soil) and the rate at which water is supplied to the freezing front (the hydrologic conditions). For example, dessication of the unfrozen soil below would lower the suction gradient, thus decreasing the flow of water. The reduced amount of latent heat liberated would allow more sensible heat to flow from the freezing front; thus the temperature would fall, r_{iw} would decrease and ice could penetrate the soil pores. Conditions suitable for ice lensing may subsequently be re-established further into the soil, resulting in the banded distribution of ice which is often observed in laboratory experiments and in nature.

An important question is: how can we specify the suction at the frost line (the so-called cryosuction)? Considering this hypothetically, one can calculate r_{iw} as a function of the temperature, from equation (5). Then using

TABLE 6-1

r_{iw} vs. TEMPERATURE

$T(°C)$	$r_{iw}\ (\mu m)$	$P_i - P_w\ (kg\ cm^{-2})$
-0.005	10.0	0.062
-0.01	5.0	0.12
-0.02	2.5	0.24
-0.05	1.0	0.62
-0.10	0.5	1.24
-0.20	0.25	2.48

TABLE 6-2

PORE SIZE CHARACTERISTICS OF IDEAL SOILS

Soil	Grain Size (μm)	Pore radius (μm)[1]
Coarse sand	500	44.6
Medium sand	375	33.5
Fine sand	150	13.4
Medium silt	30	2.7
Fine silt	10	0.89
Clay	2	0.18

[1] *Assumes that particle radius is 5.6 X pore radius*
(see Penner, 1967)

equation (4), one can calculate (P_i - P_w). These values are shown in table 6-1. If we then take an idealized soil, we can calculate a typical pore size for various soil types (table 6-2) and, by reference to table 6-1, we can estimate (P_i - P_w) for each soil. In the case of actual soils, Williams (1967) suggests that the cryosuction can be estimated from the air intrusion value in a conventional suction-moisture test. Using this, typical values are shown in table 6-3. Although clays develop high suctions, their low per-

TABLE 6-3

AIR INTRUSION VALUES FOR VARIOUS SOILS (kg cm^{-2})

Coarse sand	0
Medium and fine sand	0 - 0.075
Medium silt	0.075 - 0.15
Fine silt	0.15 - 0.5
Silty clay	0.5 - 2.0
Clays	>2.0

(From Williams, 1967)

meability restricts water flow and hence limits the rate of frost heave. Thick ice lenses are associated, in general, with high permeabilities and slow rates of freezing.

EVIDENCE OF HEAVING PRESSURES

The problem of frost heave attained considerable practical significance in the mid-1970s with plans to build refrigerated, buried gas pipelines from northern Canada. In the discontinuous permafrost zone this would cause the long term growth of permafrost *ab initio*, with the possibly serious risk of frost heaving. Early designs called for a soil berm over the pipeline in order to reduce the risk of vertical displacement. This was based on the concept of a "shut-off pressure". The argument for this is that if the weight acting on the ice, and constituting P_i, is increased by adding a load, then P_w

will be higher too by the same amount. Therefore there will be less suction effect at the freezing front and the heave will be reduced accordingly; if the load is increased sufficiently, there should be no heave at all. Based upon the capillary model, one can calculate that a soil berm of 3 m or so would be generally sufficient; at a bulk density of 0.15 kg cm^{-2} m, 3 m of overburden results in a pressure of 0.45 kg cm^{-2} (cf. table 6-3). In the worst cases, in clayey soils, one might envisage a berm of 10 m (see Williams, 1979). However, results of experiments by Penner (1967) and by Sutherland and Gaskin (1973) to measure heaving pressures had failed to corroborate the values in table 6-3, but were higher in all cases.

By the early 1970s, it was becoming clear that the capillary theory could greatly underestimate the heaving pressure, which, it is generally agreed, is derived from the difference between the pressure states of ice and water (i.e. P_i - P_w). Hoekstra (1969) obtained heaving pressures up to 5.6 kg cm^{-2} for silt and up to 24 kg cm^{-2} for clay, and the validity of these results has been confirmed by other investigators (e.g. Loch and Miller, 1975).

There is also certain field evidence which supports these findings. For example, Mackay (1971) described examples of massive segregated ice at depths exceeding 35 m in association with a variety of soils; near the base of such a lens, the overburden pressure would be in excess of 3 kg cm^{-2}, and yet the lens had continued to grow. Mackay also suggested that some pingos may be formed of segregated ice, where overburden pressures of at least 3.5 kg cm^{-2} are implied (Mackay, 1979). The presence of ice layers in permafrost at depths exceeding 30 m (Williams, 1968), presumably formed with simultaneous heaving, is further evidence of the magnitude of heaving forces. The ice was able to prise an opening beneath tens of metres of overburden.

This evidence implied that frost heaving of a buried gas pipeline would not be restrained by an earth berm of a couple of metres. Rather, the berm would have to be 20 to 30 m or more, or the pipe would have to be buried a roughly equivalent amount (Williams, 1979). There remains considerable practical, as well as scientific, interest in frost heave.

It thus appears that the capillary model concept of "ice intrusion" in freezing soils is not generally correct, and that some other mechanism must be found to explain ice segregation and high heaving pressures. The logical explanation for high heaving pressures is that the ice lens must grow at a location where the difference in pressure states between the ice and water is much greater. The relationship between heaving pressure and temperature is shown in figure 6-3; this implies that the ice lens must form at some distance behind the 0°C isotherm (Miller, 1972). Experiments by Loch and Miller (1975), Loch and Kay (1978) and Penner and Goodrich (1980) revealed that the growth of ice lenses does not usually occur at the freezing front, as indicated in the capillary model. Rather, it appears that ice lensing normally takes place in the partially frozen soil some distance behind the 0°C isotherm. One theoretical explanation for this, first proposed by Harlan (1973), has become known as the hydrodynamic model (Taylor and Luthin, 1978; Guymon et al., 1980).

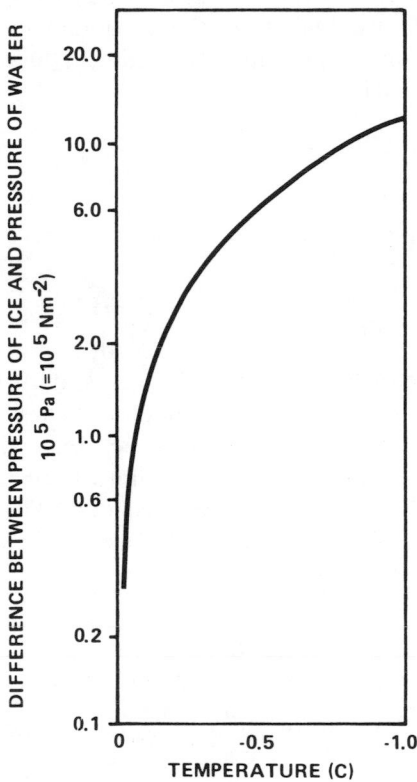

Fig. 6-3. The dependence of heaving pressure on temperature (from Williams, 1979). Note that 10^5 N/m^2 ≈ 1 kg - weight cm^{-2} = 10^2kPa

THE HYDRODYNAMIC MODEL

This approach puts the problem of soil freezing on a more general footing by recognizing the interrelatedness of heat and moisture flows in a system of coupled equations. These are adaptations of equations for heat and fluid flows in unfrozen soils, being linked by the unfrozen water content relationship (the freezing characteristic curve) and the Clapeyron equation. Typically, the system of equations is as follows.

For water flow, we have

$$\frac{\partial}{\partial z}(k\frac{\partial \psi}{\partial z}) = \frac{\partial \Theta_w}{\partial t} = \frac{\partial \Theta_u}{\partial t} + \frac{\rho_i}{\rho_w}\frac{\partial \Theta_i}{\partial t} \tag{6}$$

where k is the hydraulic conductivity, ψ is suction head (ignoring elevation head for now), Θ_w, Θ_u and Θ_i are the volumetric contents of total water, unfrozen water and ice respectively, and ρ_i is the density of ice. The freezing characteristic curve (Θ_u vs. T) dictates the maximum amount of unfrozen water that can exist at any temperature. Therefore, in (6) we can substitute

$$\frac{\partial \Theta_u}{\partial t} = \frac{\partial \Theta_u}{\partial T} \cdot \frac{\partial T}{\partial t}$$

so that

$$\frac{\partial}{\partial z}(k\frac{\partial \psi}{\partial z}) = \frac{\partial \Theta_u}{\partial T} \cdot \frac{\partial T}{\partial t} + \frac{\rho_i}{\rho_w}\frac{\partial \Theta_i}{\partial t}. \tag{7}$$

For heat flow,

$$\frac{\partial}{\partial z}(K\frac{\partial T}{\partial z}) + C_w q\frac{\partial T}{\partial z} = L\rho_i\frac{\partial \Theta_i}{\partial t} = C_m\frac{\partial T}{\partial t} \tag{8}$$

where C_w is the volumetric heat capacity of water, and q is the flux of water. On the left-hand side, the first term accounts for heat conduction due to temperature gradients, the second term is the convective heat transfer associated with the flow of water, and the third term represents a heat source (or sink) due to water-ice phase transition. The second term is typically only 0.01 to 0.001 of the first and is usually ignored (Taylor and Luthin, 1978). Equation (8) then becomes

$$\frac{\partial}{\partial z}(K\frac{\partial T}{\partial z}) + L\rho_i\frac{\partial \Theta_i}{\partial} = C_m\frac{\partial T}{\partial t} \tag{9}$$

Solving for, and equating, $\rho_i\partial \Theta_i/\partial t$ from (7) and (9), we get

$$\frac{\partial}{\partial z}(K\frac{\partial T}{\partial z} + L\rho_w\frac{\partial}{\partial z}(k\frac{\partial \psi}{\partial z}) = C_m\frac{\partial T}{\partial t} + L\rho_w\frac{\partial \Theta_u}{\partial T} \cdot \frac{\partial T}{\partial t} \tag{10}$$

From equation (2),

$$C_m + L\rho_w\frac{\partial \Theta_u}{\partial T} = C_a$$

Therefore,

$$\frac{\partial}{\partial z}(K\frac{\partial T}{\partial z}) + L\rho_w\frac{\partial}{\partial z}(k\frac{\partial \psi}{\partial z}) = C_a\frac{\partial T}{\partial t} \tag{11}$$

(In the case of no water flow, equation (11) reduces to equation (1)). On the left-hand side, the second term accounts for the heat generated by the formation of segregated ice, whilst the right-hand side includes the heat generated as interstitial water freezes *in situ*.

The hydrologic state is seen as continuous from unfrozen into frozen soil. In the unfrozen soil, ψ depends on Θ_w. In the frozen soil, gradients of soil water potential (suction gradients) are associated with water coexisting with ice at temperatures below 0°C. The potential of the unfrozen water is defined by the Clapeyron equation,

$$\psi(T) = L(T - T_o)/V_w T_o \qquad (12)$$

The rate of water flow depends, in general, on the gradient of potential and the hydraulic conductivity, $k(\Theta_w)$. In frozen soil, water flow depends, then, on the temperature gradient and on temperature itself; the latter affects the unfrozen water content and hence the conductivity (see below). This relationship varies, of course, with soil type.

Guymon *et al.* (1980) discussed various strategies for solving the system of equations. A heat balance accounting is applied at the freezing front, where the net heat flow,

$$Q = K_f \left(\frac{\partial T}{\partial z}\right)_f - K_u \left(\frac{\partial T}{\partial z}\right)_u \qquad (13)$$

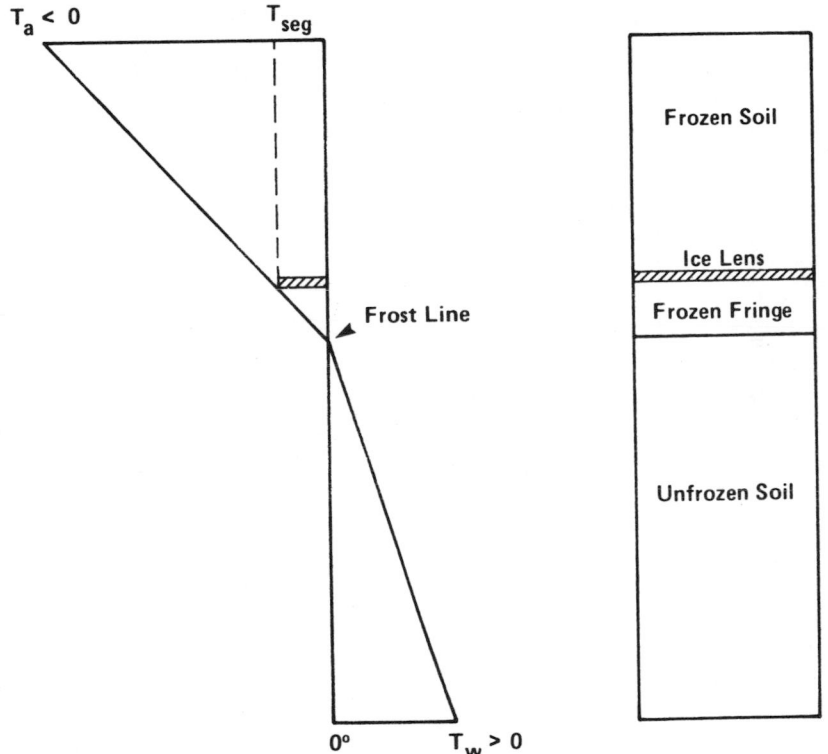

Fig. 6-4. Temperature conditions associated with ice lensing (from Konrad and Morgenstern, 1980)

is partitioned between the freezing of interstitial water, the freezing of any water which has migrated there, and the net cooling of the soil (see also Holden *et al.*, 1980). This determines whether ice accumulation and frost heave can continue or not.

To return to the phenomenon of ice segregation itself, we have a view in the hydrodynamic model of the ice lens growing within the frozen soil some distance behind the freezing front (figure 6-4). In order for the ice lens to grow, water migration must occur through the intermediate partially frozen zone known as the "frozen fringe" (Miller, 1972). That water can and

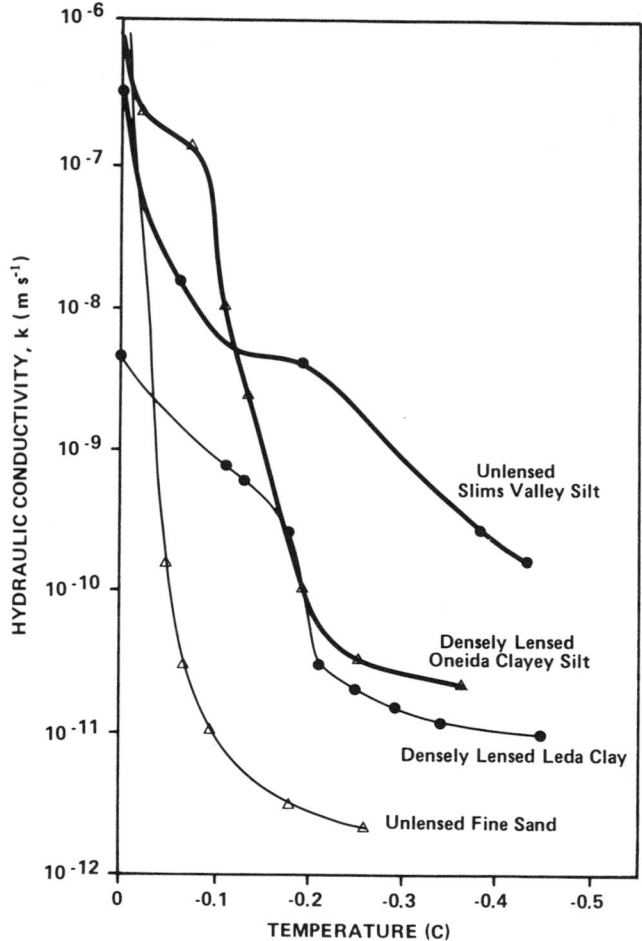

Fig. 6-5. Hydraulic conductivity for various frozen soils (from Burt and Williams, 1976)

indeed does flow through frozen soil was demonstrated by Hoekstra (1969). The first direct determinations of frozen ground permeabilities were made by Burt and Williams (1976); the values are low (see figure 6-5), but should be viewed together with the very considerable potential gradients which are developed in freezing soils. For a temperature difference between 0° and -0.5°C, for example, the suction difference is 6 kg cm^{-2} (about 6 atmospheres); at -1°C, the suction is 12 kg cm^{-2} (figure 6-3). (Even so, the values reported by Burt and Williams may, in fact, be too high. Values reported by Loch and Kay (1978) and by Konrad and Morgenstern (1980) are about two orders of magnitude smaller).

The explanation of ice segregation in freezing soil according to the hydrodynamic model is quite different from that according to the capillary model. If we imagine water being drawn towards and into freezing soil, we encounter a very sharp decrease in hydraulic conductivity within the frozen soil. If, for simplicity, we assume a linear temperature gradient (i.e., a linear gradient of potential), then there will be a divergence of water flux, since $q_u >> q_f$ where u and f refer to unfrozen and frozen zones respectively. Thus, water will accumulate at some point and freeze, creating an ice lens. The zone of ice accumulation corresponds to the point of maximum flow divergence, i.e. where dk/dT is maximum. From figure 6-5 we see that this occurs in the temperature range -0.2 to -0.3°C or so. This in turn corresponds to the temperature at which $d\Theta_u/dT$ is maximum (cf. figure 6-1). Since at 0°C, k is the same as in the unfrozen soil, this explains why the ice lens grows somewhere behind the freezing front. The picture just described is summarized diagrammatically in figure 6-6.

Conditions for the initiation and termination of ice lens growth according to the hydrodynamic model may be outlined only in a rather general way (Konrad and Morgenstern, 1980). When a soil is subjected to a step-change freezing temperature, then initially the rate of frost penetration is high. The net heat flow at the freezing front is so great that water cannot migrate at a sufficient rate to create a continuous ice lens at a given level, although some ice enrichment may occur. However, as the rate of freezing slows, thin rhythmic lenses will appear (figure 6-7). The view of Konrad and Morgenstern (1980) is that, under conditions of unsteady heat flow, the freezing front must continue to advance into the soil; this causes the frozen fringe to increase in thickness and the temperature drop across it to become greater. Both of these factors cause a decrease in the overall permeability of the fringe (at a faster rate than the increase in the suction gradient), the temporary balance between heat flow and water supply is broken, and ice lens growth stops in that location. A new, lower level of ice accumulation is established in accordance with the local permeability. Thus, under unsteady conditions, it is the transmission of water to the growing ice lens that

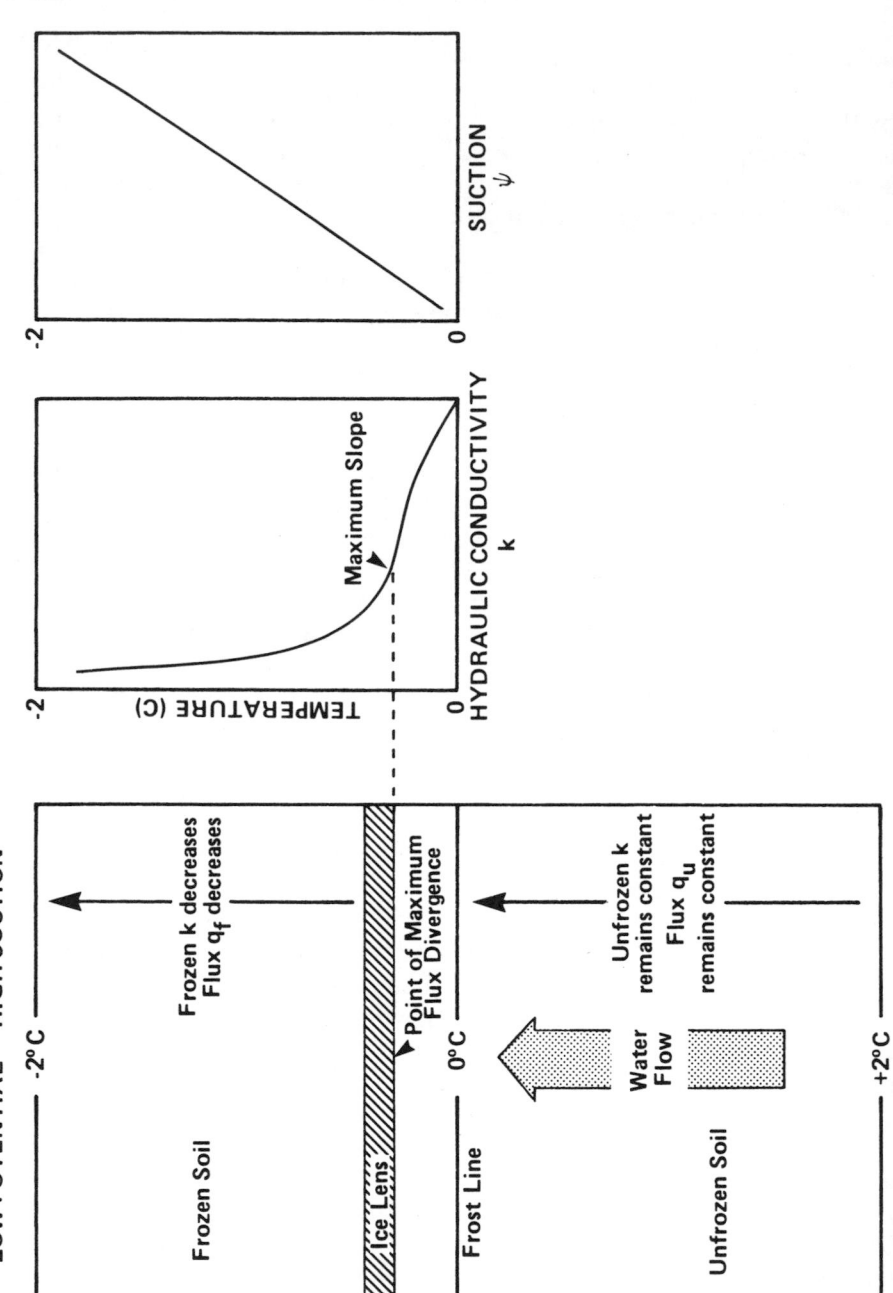

Fig. 6-6. Ice lensing according to the hydrodynamic model (from Williams and Wood, 1982)

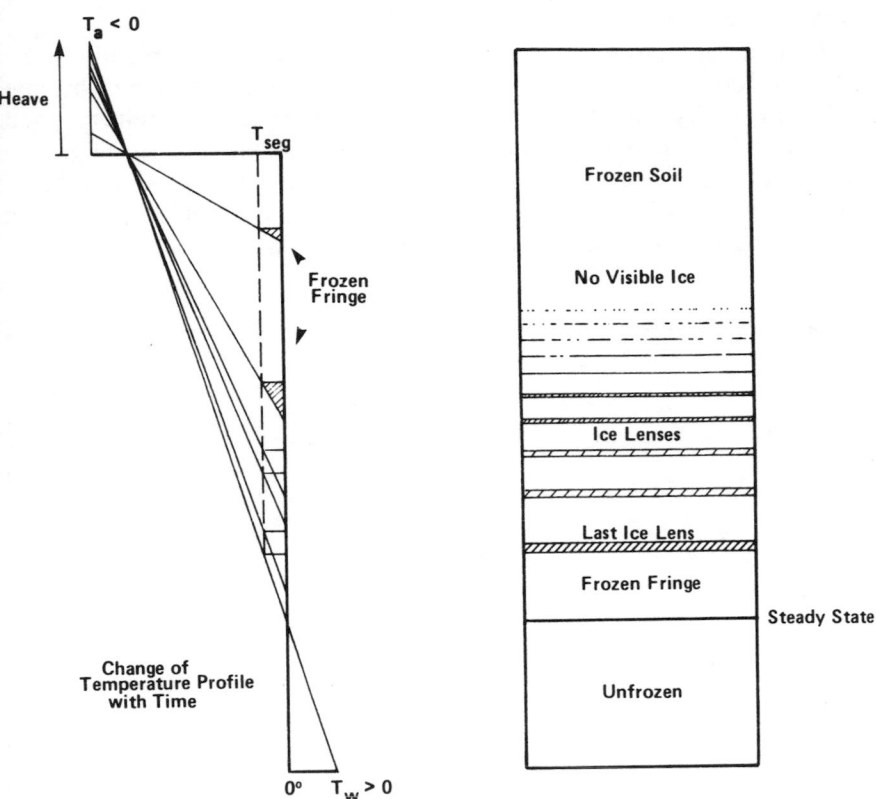

Fig. 6-7. Schematic rhythmic ice lens formation (from Konrad and Morgenstern, 1980)

is the limiting condition (i.e. a hydraulic condition), the frost line advancing with successive ice lenses, which get thicker with depth (figure 6-7). This has been termed secondary heaving by Miller (1972).

Primary heaving occurs when the frost line is stationary and the net heat flow just balances the latent heat generated by the freezing water, i.e.,

$$Q = L \cdot k \partial \psi / \partial z \qquad (14)$$

(cf. Arakawa, 1966). This is the form of ice segregation envisaged in the capillary theory; however, it is probably only rarely achieved in nature for any length of time. This is because either water flow will usually diminish with dessication of the unfrozen soil below, and/or the net heat flow will diminish as the ice lens gets thicker. Mackay (1971) has explained the formation of massive ice segregation in terms of primary heaving with pore water replenishment from below (pp. 412-415).

With the penetration of frost into a soil, the heaving rate is initially small, as water flow is insufficient to balance the net heat flow and the frost line advances rapidly into the soil. With time, the net heat flow diminishes and the heaving rate increases to a maximum, when conditions come closest to those specified in (14). Beyond this, net heat flow continues to diminish as a thermal steady state is approached ($Q=0$), so that eventually the heaving rate becomes zero. These features are illustrated in figure 6-8.

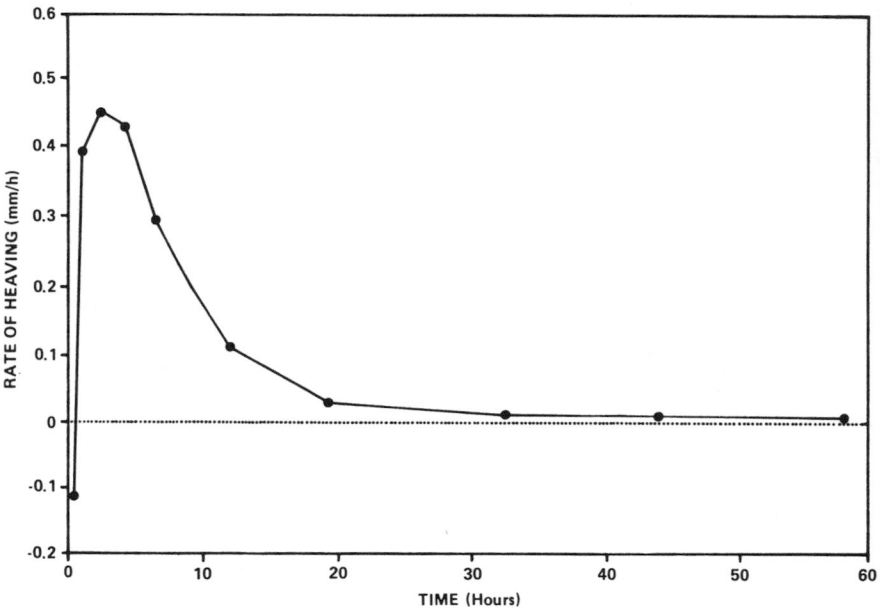

Fig. 6-8. Heave rate versus time for a compacted till (from Garand, 1981)

The accuracy with which the hydrodynamic model will predict water redistribution and frost heaving is not yet fully established, largely because of the general lack of comprehensive test data. However, some limited evaluations are available and show encouraging results. Between them, various versions are able to reproduce the range of phenomena described above, qualitatively at least. In a quantitative way, Taylor and Luthin (1978) reported some success in prediction of total water content and heave. Guymon et al. (1980) predicted heave with excellent results; Holden et al.

(1980) consistently overpredicted heave by about 25 percent. However, all authors agree that model predictions are highly sensitive to hydraulic parameters and boundary conditions, which seems to imply that such an approach is unlikely to lead to a *practical* frost heave model (Konrad and Morgenstern, 1980). This matter is taken up in the final section.

HEAVING WITHIN FROZEN GROUND

Although the hydrodynamic model has been applied only to water redistribution in the vicinity of a freezing front, there is no reason to suppose that migration of water and accumulation of ice in existing frozen ground should be any different from that already described. Because of the continuity and mobility of water within the network of unfrozen water interfaces, water migration, ice lensing and frost heave may continue, albeit slowly, within permafrost long after any primary or secondary heaving has ceased. Miller (1972) has termed this tertiary heaving.

Although the permeability of frozen ground is very low, enormous suction gradients are associated with temperature gradients and significant flow could occur over the long term. In fact, Harlan (1974) suggested that such a process could explain the widespread occurrence of ice-rich layers at the top of permafrost. Normal mean temperature gradients imply long term water movement from depth (warmer) towards the surface (colder), to accumulate as ice. Water redistribution within frozen soil has been observed in laboratory experiments by Mageau and Morgenstern (1980) and by Penner and Goodrich (1980) (figure 6-9). Actual field observations to date are limited to those of Mackay *et al.* (1979), who described frost heave of 1 to 2 cm in apparently frozen ground between January and May. Such results are obviously of much interest for the design of refrigerated structures in permafrost.

THE SECONDARY FROST HEAVE MODEL

Miller (1972) distinguished three types of frost heaving:

i) primary, where the base of the ice lens coincides with the frost line;
ii) secondary, where ice lenses form behind a frozen fringe (associated with unsteady conditions);
iii) tertiary, referring to water redistribution within existing frozen ground.

It appears that primary heaving would occur very rarely in nature, since it requires steady conditions, and that secondary heaving is the common

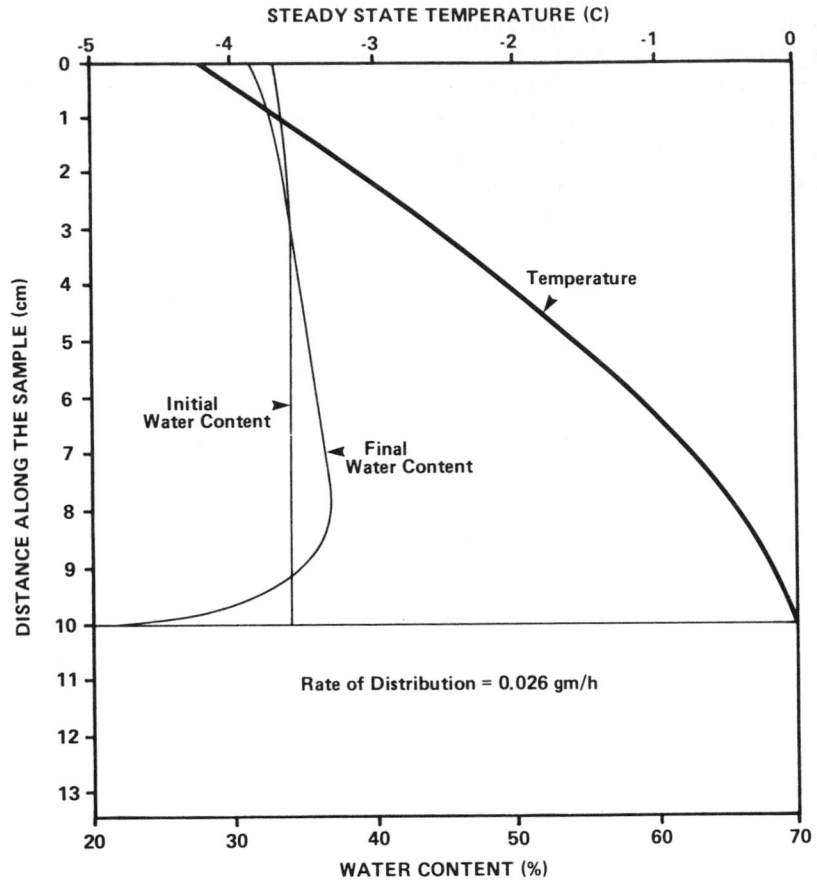

Fig. 6-9. Moisture migration in a 10 cm sample of pre-frozen Devon silt during closed system freezing (time of experiment = 16 days) (from Mageau and Morgenstern, 1980)

form. Further, secondary heaving produces larger heaving pressures (since the base of the ice lens is at a lower temperature), and can explain the high pressures observed in experiments and inferred from field observations.

Miller (1972) reasoned that, during the formation of pore ice in the frozen fringe, the pressure P_i would not be constant and equal to the overburden pressure (as assumed in the capillary model), but that there is an ice pressure gradient, as well as a water pressure gradient (figure 6-10a). The generalized form of the Clapeyron equation applies:

$$V_iP_i - V_wP_w = L\Delta T/T_o \tag{15}$$

Therefore

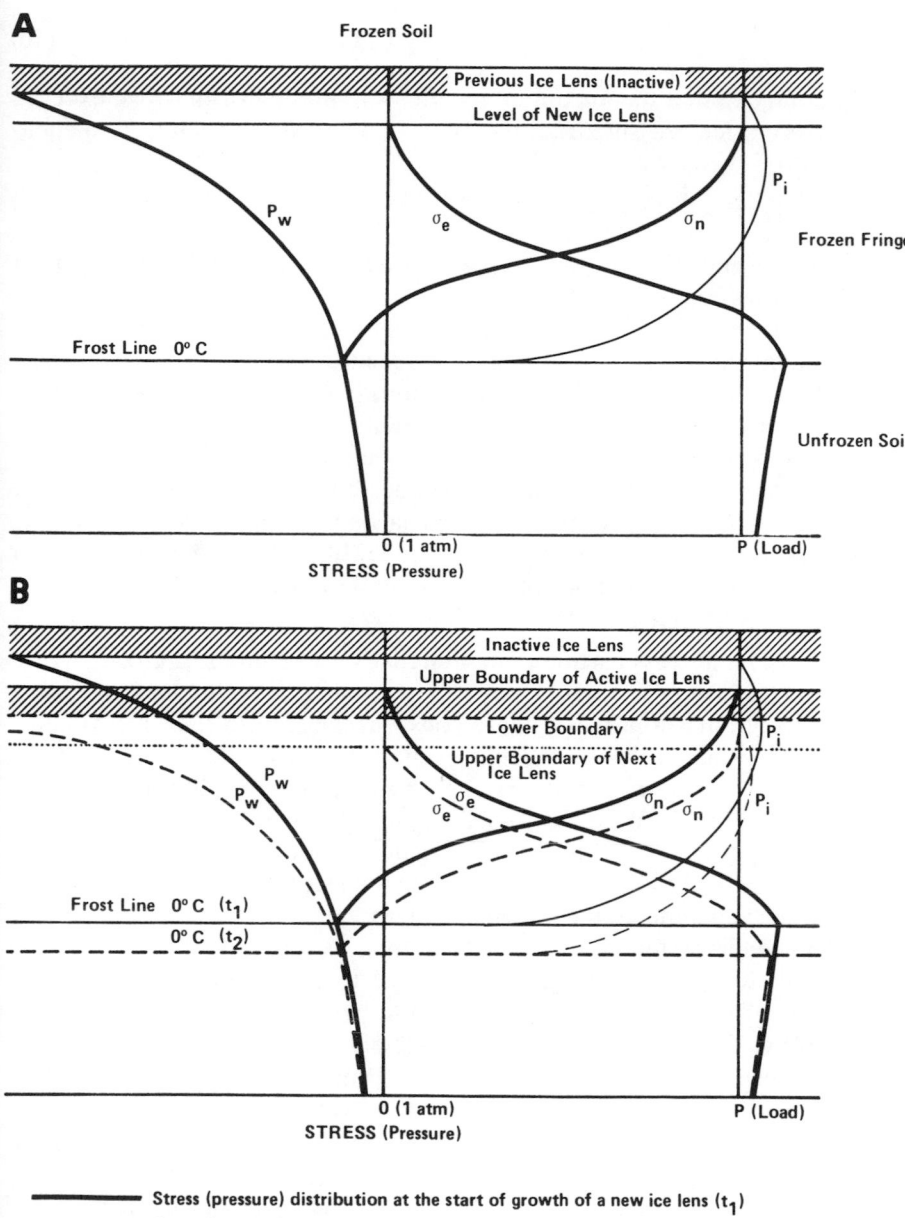

Fig. 6-10. Stress and pressure distributions in freezing soil: a) at the start of a new ice lens, b) when the ice lens stops growing (adapted from Williams and Wood, 1982). The soil is saturated, but pore water pressure at and below the frost line is assumed to be pulled below atmospheric pressure as the result of the very large suctions in the freezing soil.

$$\frac{\partial P_i}{\partial z} = \frac{1}{V_i}(V_w\frac{\partial P_w}{\partial z} + \frac{L}{T_o}\frac{\partial}{\partial z}(\Delta T)) \tag{16}$$

Miller argued that the maximum ice pressure in the frozen fringe exceeds the overburden pressure (figure 6-10a). The heaving pressure, from equation (15), is given by

$$P_i = \frac{1}{V_i}(V_wP_{wb} + L\Delta T_b/T_o) \tag{17}$$

where P_{wb} and ΔT_b are the unfrozen water pressure and freezing point depression at the base of the ice lens.

Miller employed the same notions of heat and water flow as the hydrodynamic model, but also included ice transport in the frozen fringe. In addition, he also provided specific criteria for locating an ice lens behind the freezing front, based on stress conditions. This allows us to specify the conditions for ice lens initiation and termination.

The load of the soil, P, is partially supported by the reaction stress of the soil matrix (the effective stress, σ_e) and partially by the reaction of the pore contents (the neutral or buoyant stress, σ_n) (Miller, 1978). When pore water and pore ice are continuous phases, Miller adopts the relationship

$$\sigma_n = P_i - \chi(\zeta)(P_w - P_i)$$

where $\chi(\zeta)$ is a stress partition function that varies with the unfrozen water content (ζ is a temperature-dependent parameter). In the frozen fringe, pore ice moves from warmer to colder regions via regelation (Miller, 1972; Philip, 1980): the soil particles will remain stationary as long as the load, P, exceeds the buoyant stress, σ_n (i.e., as long as $\sigma_e > 0$). However, σ_n increases throughout the frozen fringe with P_i: where $\sigma_n = P$, $\sigma_e = 0$, and soil particles are no longer pressed against stationary neighbours below and will be pushed apart by the moving ice (Miller, 1978). The widening gap will accommodate the growing ice lens.

The condition for ice lens initiation, then, is that $\sigma_e = 0$ (figure 6-10a). The lens will continue to grow as long as $P_i > P$. However, as the lens continues to grow, P_w falls as the soil below becomes desiccated (assuming that water flow is insufficient to completely replenish the soil pores). In accordance with the Clapeyron equation, P_i must then fall also; when $P_i \leqslant$ P, the ice lens can no longer displace the overlying soil and it ceases to grow at that point. A new lens then forms below, where $\sigma_e = 0$ (figure 6-10b). The process repeats in a rhythmic fashion.

Miller (1978) stated that the secondary heave model appears to be amenable to computer simulation, although adequate comprehensive data are not yet available and solutions for transient boundary conditions will be difficult (Miller and Koslow, 1980).

CONCLUDING REMARKS

The presence of a continuous, unfrozen water phase separating ice from the mineral matrix in "frozen" soils is accepted. Further, the factors that govern the amount and mobility of this water are now understood, at least qualitatively. However, there is still some disagreement over the precise, quantitative description of ice segregation, although the significance of a frozen fringe has been established.

This paper has reviewed three models of ice lensing in freezing soils: the capillary, hydrodynamic and secondary heaving models. Of these, the capillary model is unique in visualizing the base of an ice lens coincident with the frost line. However, such a situation rarely, if ever, actually occurs. The hydrodynamic and the secondary heaving models visualize the growing ice lens at some distance behind the frost line, separated from it by a frozen fringe. Accordingly, both models are able to account for the high heaving pressures measured in experiments or inferred from field observations. These models share notions concerning the coupled flow of heat and water, but the secondary model is the most comprehensive since it includes the consideration of stress conditions and the effects of overburden pressure.

A comprehensive theory of frost heaving should predict the rate of frost penetration, rate of heaving and heaving pressures from a set of equations describing the thermal, hydrologic and stress conditions in the soil. Only very limited testing of proposed models has taken place because of the unavailability of input data. While the Clapeyron equation can be used for $\psi(T)$, realistic and representative data for $K(T)$, $C_a(T)$, $\Theta_u(T)$, $k(T)$, etc. are not generally available. These properties vary considerably between soils and must be determined by experiment. Further, model predictions have been shown to be highly sensitive to changes in input parameters; because of this, and the considerable variation of soil conditions in nature, the application of these models to real world problems will be difficult, if not impossible.

While the search for a rational theory of frost heaving is an important scientific endeavour, it offers no immediate prospect for solving important practical problems of construction in cold regions. From an applied point of view, predicting the location and thickness of individual lenses is not so important, although an understanding of the factors influencing the rate of heaving is required to assess whether or not such behaviour is acceptable when freezing occurs over some finite period of time (e.g., 30 years). In this regard, the idea of using an earth berm to introduce some shut-off pressure and thus prevent any heaving of a buried, refrigerated pipeline is now understood to be impractical. However, some control of the heave rate is possible by applying a surcharge on the pipe (either with a berm or by deeper burial). In any event, some estimate of heave rate or total heave is required for the design of refrigerated structures.

Hwang and Yip (1977) proposed an upper bound solution for the frost heave of a buried pipe as a design method from a practical point of view. The maximum possible heave occurs when the net heat flow just balances the latent heat generated by the freezing of migrating water (equation 14). In this case, the heave rate, dH/dt, equals the frost penetration rate, dZ_f/dt. Based upon this, one would seek to limit dZ_f/dt, by the use of insulation for example, thus keeping the total heave over the life of the structure within acceptable limits. Under actual field conditions, though, water may not migrate fast enough to form pure segregated ice (as in equation (14)), and thus a somewhat more refined approach is desirable. Laboratory tests could be carried out to determine dH/dt over a range of freezing rates for any variety of soils. (Hydrologic conditions could be standardized, with all tests run at saturation.) A thermal model could be used to estimate dZ_f/dt at various time intervals, and then based on the results from the laboratory tests, one could estimate dH/dt. Since thermal models, *per se*, seem inherently less sensitive than coupled flow models, the prediction of dZ_f/dt could be made fairly reliable. In addition, the variation of soil thermal properties in nature is much less than the variation of hydraulic properties. Hence, from the practical point of view, the approach outlined would seem to be a viable one.

ACKNOWLEDGEMENTS

I would like to acknowledge the many useful discussions I have had with colleagues and students in the Geotechnical Science Laboratories, Carleton University. In particular, I wish to express my thanks to Peter Williams, D.E. Patterson and J.A. Wood.

REFERENCES

Arakawa, K. 1966. Theoretical studies of ice segregation in soil. J. Glaciology 6: 255-260.
Burt, T.P. and Williams, P.J. 1976. Hydraulic conductivity in frozen soils. Earth Surface Processes 1: 349-360.
Everett, D.H. 1961. The thermodynamics of frost damage to porous solids. Faraday Soc., Trans. 57 (465, Part 9): 1541-1551.

Garand, P. 1981. Méthodologie experimentale permettant l'étude de la gelivité d'un till en fonction du mode de compactage. M.Sc.A. Thesis, Univ. Montreal, Dept. Civil Eng.: 179 pp.

Gold, L.W. and Lachenbruch, A.H. 1973. Thermal conditions in permafrost: a review of North American literature. Permafrost 2nd Int. Conf., Yakutsk, U.S.S.R., July 13-28: North American Contribution. Washington, D.C., NAS (U.S.A.): 3-25.

Goodrich, L.E. and Gold, L.W. 1981. Ground thermal analysis. *In* Johnston, G.H. ed., Permafrost: Engineering design and construction. N.Y., John Wiley and Sons: 149-172.

Guymon, G.L., Hromadka, T.V. II and Berg, R.L. 1980. A one-dimensional frost heave model based upon simultaneous heat and water flux. Cold Regions Science and Technology *3*: 253-262.

Harlan, R.L. 1973. Analysis of coupled heat-fluid transport in partially frozen soil. Water Resources Research *9*: 1314-1323.

Harlan, R.L. 1974. Dynamics of water movement in permafrost: a review. *In* Permafrost Hydrology, Workshop/Seminar, Calgary, Canada, February 18-20. Can. Nat. Comm., Int. Hydrological Decade: 69-78.

Harlan, R.L. and Nixon, J.F. 1978. Ground thermal regime. *In* Andersland, O.B. and Anderson, D.M. eds., Geotechnical engineering for cold regions. N.Y., McGraw-Hill: 103-163.

Hoekstra, P. 1969. Water movement and freezing pressures. Soil Science Soc. Amer. Proc. *33*: 512-518.

Holden, J.T., Jones, R.H. and Dudek, S.J-M. 1980. Heat and mass flow associated with a freezing front. Presented at 2nd Int. Symp. on Ground Freezing, Trondheim, Norway, June 24-26. Preprints: 502-514.

Hwang, C.T. and Yip, F.C. 1977. Advances in frost heave prediction and mitigation methods for pipeline application. Amer. Soc. Mech. Engineers, Paper #77-WA/HT-19: 11 pp.

Konrad, J.M. and Morgenstern, N.R. 1980. A mechanistic theory of ice lens formation in fine-grained soils. Can. Geotech. J. *17*: 473-486.

Loch, J.P.G. and Kay, B.D. 1978. Water redistribution in partially frozen, saturated silt under several temperature gradients and overburden loads. Soil Science Soc. Amer. Proc. *42*: 400-406.

Loch, J.P.G. and Miller, R.D. 1975. Tests of the concept of secondary heaving. Soil Science Soc. Amer. Proc. *39*: 1036-1041.

Mackay, J.R. 1971. The origin of massive icy beds in permafrost, Western Arctic coast, Canada. Can. J. Earth Sciences *8*: 397-422.

Mackay, J.R. 1979. Pingos of the Tuktoyaktuk Peninsula area, Northwest Territories. Géographie Phys. Quat. *33*: 3-61.

Mackay, J.R., Ostrick, J., Lewis, C.P. and Mackay, D.K. 1979. Frost heave at ground temperatures below 0°C, Inuvik, Northwest Territories. Geol. Surv. Canada Paper *79-1A*:403-405.

Mageau, D.W. and Morgenstern, N.R. 1980. Observations on moisture migration in frozen soils. Can. Geotech. J. *17*: 54-60.

Miller, R.D. 1972. Freezing and heaving of saturated and unsaturated soils. Highway Research Record *393*: 1-11.

Miller, R.D. 1978. Frost heaving in non-colloidal soils. 3rd Int. Conf. on Permafrost, Edmonton, Alberta, July 10-13. Proc. *1*. Ottawa, NRC Canada Pub. *16529*: 707-713.

Miller, R.D. and Koslow, E.E. 1980. Computation of rate of heave versus load under quasi-steady state. Cold Regions Science and Technology *3*: 243-252.

Patterson, D.E. and Smith, M.W. 1981. The measurement of unfrozen water content by time domain reflectometry: results from laboratory tests. Can. Geotech. J. *18*: 131-144.

Penner, E. 1959. The mechanism of frost heaving in soils. Highway Res. Bd., Washington, Bull. *225*. NAS/NRC(USA), Pub. *685*: 1-22.

Penner, E. 1967. Heaving pressure in soils during unidirectional freezing. Can. Geotech. J. *4*: 398-408.

Penner, E. and L.E. Goodrich. 1980. Location of segregated ice in frost susceptible soil. Presented at 2nd Int. Symp. on Ground Freezing, Trondheim, Norway, June 24-26. Preprints: 626-639.

Philip, J.R. 1980. Thermal fields during regetation. Cold Regions Science and Technology *3*: 193-204.

Riseborough, D.W., Smith, M.W. and Halliwell, D.H. 1983. Determination of thermal properties of frozen soils. 4th Int. Conf. on Permafrost, Fairbanks, Alaska, July 18-22. Proc. NAS (USA), National Academy Press: 1072-1077.

Sutherland, H.B. and Gaskin, P.N. 1973. Pore water and heaving pressures developed in partially frozen soils. Permafrost 2nd Int. Conf., Yakutsk, U.S.S.R., July 13-28: North American Contribution. Washington, D.C., NAS(U.S.A.): 409-419.

Taylor, G.S. and Luthin, J.N. 1978. A model for coupled heat and moisture transfer during soil freezing. Can. Geotech. J. *15*: 548-555.

Williams, P.J. 1967. Properties and behaviour of freezing soils. Norwegian Geotech. Inst., Oslo. Pub. *72*: 119 pp.

Williams, P.J. 1968. Ice distribution in permafrost profiles. Can. J. Earth Sciences *5*: 1381-1386.

Williams, P.J. 1979. Pipelines and Permafrost. London, Longman: 98 pp.

Williams, P.J. and Wood, J.A. 1982. Investigation of moisture movements and stresses in frozen soils. Canada Dept. Energy, Mines and Resources, Earth Physics Br., Contract OSU81-00119. Final Rep.: 151 pp.

7

A STEP FUNCTION MODEL
OF ICE SEGREGATION

S.I. Outcalt

Department of Geology, The University of Michigan
Ann Arbor, MI 48104 U.S.A.

ABSTRACT

Using extremely simple algebraic expressions a step function model has been developed for segregation frost. The model yields estimates of the depth of the normally frozen layer, the duration of normal frost, the total heave and the duration of segregation frost. Surface temperature depressions of -2 to -20°C and water table depths between 10 and 100 cm were used to simulate open system freezing. Ice lens thickness was found to vary between 18 and 512 cm and the duration of lens growth to range from 2 to 922 years depending upon initial conditions.

RÉSUMÉ

Un modèle de ségrégation de glace à base d'une fonction en échelon

Un modèle de ségrégation de la glace a été developpé sur la base d'une fonction algébraique très simple. Le modèle permet d'estimer la profondeur de la couche normalement congélée, la durée du gel normal, le soulèvement total, et la durée du gel de ségrégation. Les abaissements de la température de surface, soit de -2 à -20°C, et les niveaux de la nappe phréatique entre -10 et -100 cm sont utilisés pour simuler la congélation en système ouverte. L'épaisseur des lentilles de glace variaient de 18 à 512 cm et la durée d'accroissement des lentilles s'étendait de 2 jusqu'à 922 années, selon les conditions initiales.

МОДЕЛЬ В ФОРМЕ СТУПЕНЧАТОЙ ФУНКЦИИ, ОПИСЫВАЯ ЛЬДОВЫДЕЛЕНИЯ

С. И. АУТКАЛТ

РЕЗЮМЕ

Употребляя очень простые алгебраические выражения, изготовлена модель для изображения льдовыделения. Модель дает оценку глубины нормально замерзшего слоя, продолжительность нормального замерзания, полное пучение и продолжительность льдовыделения. Для симулирования замерзания открытой системы применялись температурные депрессии на поверхности от -2 до -20° и глубины водного уровня от 10 до 100 см. Мощность ледяных линз колебалась от 18 до 512 см и продолжительность роста линз изменялась от 2 до 922 лет, в зависимости от исходных условий.

INTRODUCTION

After nearly a decade of modeling ice segregation processes I found myself before a graduate seminar remarking that it was "perfectly clear" how ice segregation would occur as the product of the transport and property coupling equations I had placed on the blackboard. A perceptive student remarked that he could not see how the system worked with all the feedback within the non-linear, coupled transport equations. It was then perfectly clear to me that he was correct. Recalling Professor Mackay's maxim that "simple is best," I therefore constructed the following model which uses a simple step boundary condition in the same manner as the Terzaghi equation for dry frost and the Stefan equation for wet frost with extremely simple material property specifications. The model is intended primarily as an instructional device, but it may very well be of utility in making estimates of phenomena at geomorphic scale and frequency.

Step functions in the regime of surface temperature are commonly employed to estimate the rate of frost penetration into soils. These solutions require the thermal conductivities of unfrozen and frozen soils (K_u, K_f), the volumetric heat capacities of these materials, (C_u, C_f) and the surface temperature (T_a). The solution applied to dry soils is that of Terzaghi (1952). The material is assumed to be at 0°C at all depths as an initial condition, followed by an instantaneous surface temperature depression to T_a for all future time. This surface "step boundary condition" is the same for all the models discussed in this paper. The depth of frost penetration (Z_f) is estimated by equation (1). No subscripts are used with the thermal properties as these are only weak functions of temperature in dry soils. Here t is time in appropriate units.

$$Z_f = \sqrt{[12\,(K/C)t]} \tag{1}$$

Note that the depth of frost penetration (Z_f) is controlled only by the thermal diffusivity (K/C) and time (t), being independent of the magnitude of the surface thermal disturbance.

Terzaghi (1952) modified the equation developed by Stefan (1891) for wet soils, using saturated pore space (V_p), the latent heat of fusion (L) and a term to account for the undercooling of the frozen layer ($C_f(|T_a|/2)$). The resulting equation is

$$Z_f = \sqrt{[2K_f|T_a|t)/(LV_p + C_f\,(|T_a|/2)]} \tag{2}$$

Note that in this "wet soil case" the depth of frost penetration is dependent upon time, material properties and the magnitude of the thermal disturbance at the surface. The influence of the heat flux from the unfrozen zone is neglected to achieve a simple algebraic solution by integration:

$$\int_0^{Z_f} ZdZ = \int_0^t [(K_f|T_a|)/(LV_p + C_f\,(|T_a|/2))]dt \tag{3}$$

A slightly more complex expression was developed by Neumann which includes the effects of heat flux from the unfrozen zone.

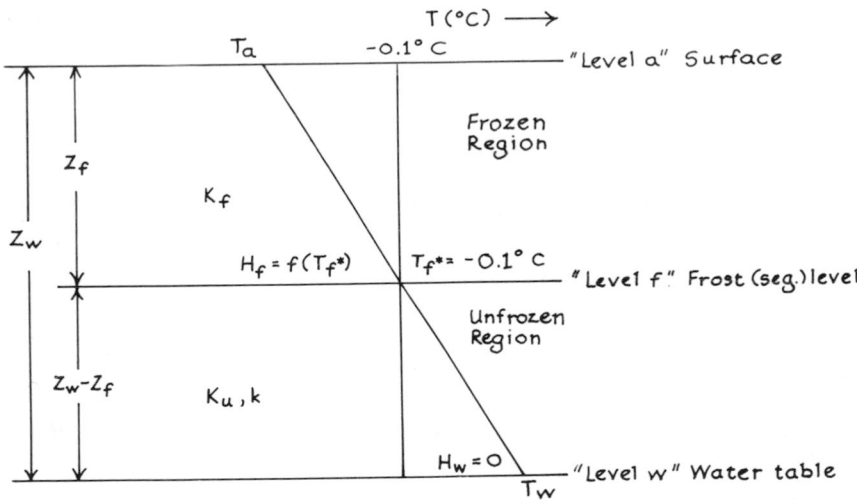

Fig. 7-1. Geometry at onset of lens growth.

Soil heave in this "wet case" is limited to $((\rho_w/\rho_i) - \rho_w) \, V_p \, Z_f$ where (ρ_w/ρ_i) is the ratio of water/ice densities. The model assumes that no ice segregation occurs and that heave is produced by the volumetric expansion of saturated pore space on freezing. Thus, an upper limit for heave produced by normal frost (without segregation) can be estimated for field situations where the depth of frost and soil porosity are known. This permits the identification of segregation effects from the measurement of surface heave, frost depth and porosity at field sites.

Next, consider the case of "wet frost" with ice segregation when it occurs over a stable water table. The lower boundary condition is a water table at a fixed above-freezing temperature. The system is considered to be "open." Water removed from the water table does not produce a drop in the elevation of the water table. The geometry of this system is illustrated as figure 7-1. Arakawa (1966) developed a segregation index based on estimated heat and water flux in a freezing soil system. Here k and ψ_t are employed for hydraulic conductivity and soil water potential. The formulation of the Arakawa model is as follows:

$$Q_f = K_f(\partial T/\partial Z)_f: \qquad \text{heat flux in the frozen zone;} \qquad (4.1)$$

$$Q_u = K_u(\partial T/\partial Z)_u: \qquad \text{heat flux in the unfrozen zone;} \qquad (4.2)$$

$$Q_w = Lk(\partial \psi_t/\partial Z_u): \qquad \text{heat loss required to freeze water} \qquad (4.3)$$
$$\text{arriving in the freezing zone from}$$
$$\text{the water table;}$$

$$\text{div} Q = Q_f - Q_u: \qquad \text{heat flux divergence in the} \qquad (4.4)$$
$$\text{freezing zone;}$$

$$\text{if div} Q > Q_w: \qquad \text{normal frost occurs;} \qquad (4.5)$$

$$\text{if div} Q \leqslant Q_w: \qquad \text{segregation frost occurs.} \qquad (4.6)$$

The segregation index can be calculated as the ratio

$$F = Q_w/\text{div} Q : \text{if } F \geqslant 1, \text{ ice lensing will occur.} \qquad (4.7)$$

To review, there are three step function models of soil frost development. These are the Terzaghi, Stefan and Arakawa models and cover the "dry", "wet" and "wet-segregation" cases. The Terzaghi and Stefan models are currently employed in engineering practice. It is the aim of this paper to produce an extremely simple step function model of ice segregation to obtain an abstract understanding of the space-time dependent behaviour of the ice lensing system. Earlier, Palmer (1967) produced a simple analytical solution to the ice lensing problem. The problem will be considered as being composed of two distinct parts. First, the evolution of the system from the time of surface freezing to the onset of lensing will be considered, followed by an analysis of further development with ice segregation active.

SYSTEM EVOLUTION WITH NORMAL FROST

The energy budget geometry illustrated in figure 7-1 is similar to that of the Stefan model with the addition of water flux. At the instant when ice segregation begins, the Arakawa index must be equal to unity. To produce a finite difference form of the Arakawa model it is necessary to specify the temperature and soil water potential (T_{f^*}, ψ_{tf^*}) at the location where water is freezing, as well as the controlling hydraulic conductivity (k) below that region. An expedient approach is to assume that ice formation occurs at some subfreezing temperature. Here the temperature of -0.1°C has been somewhat arbitrarily selected. As there is a presumed relationship between soil water potential and temperature in freezing soils ($\partial\psi_t/\partial T = 12.43 \times 10^3$ cm $H_2O/°C$), ψ_t (T) and k(T) can be specified. For typical sandy loams, reasonable values are -1243 cm H_2O and 1×10^{-7} cm sec^{-1}. The hydraulic conductivity is assumed to be the value active in the region between the 0°C isotherm and the freezing isotherm (-0.1°C). In the initially saturated system of medium textured soils, this is probably the highest soil water tension encountered in the unfrozen zone during the evolution of soil frost and thus controls the soil water potential gradient. The gravity potential is relatively small and therefore not considered in what follows. Bulk thermal conductivities of the materials were estimated using the assumptions of a porosity of 38% and total frost and saturation in the frozen and unfrozen regions. These assumptions, whilst rather sweeping, permit the Arakawa model to be expressed in a transcendental form:

$$K_f\frac{T_{f^*} - T_a}{Z_f} - K_u\frac{T_w - T_{f^*}}{Z_w - Z_f} - Lk\frac{\psi_w - \psi_{f^*}}{Z_w - Z_f} = f(Z_f) \tag{5}$$

where $f(Z_f) \to 0.0$.

Using the Newton method to optimize estimates of Z_f, convergence to within 1×10^{-10} cal cm^{-2} sec^{-1} was achieved in under ten iterations using initial values for Z_f of 0.2 cm and ($Z_w - 0.2$ cm).

SYSTEM EVOLUTION AFTER SEGREGATION

The geometry of the soil frost after segregation is shown in figure 7-2. It is assumed that the thickness of the unfrozen layer (Z_u) and the normally frozen layer (Z_f) remain constant as growth continues. All vertical expansion is due to the heave of the ice lens ($Z_i = H$). In an "open system" it is further assumed that as Z_u remains constant there will always be sufficient water for lens growth. This occurs because continued heave (increasing Z_i) will diminish the heat flux to the surface and the segregation index ($Q_w/divQ$) will be forced to values increasingly greater than unity as Q_w and

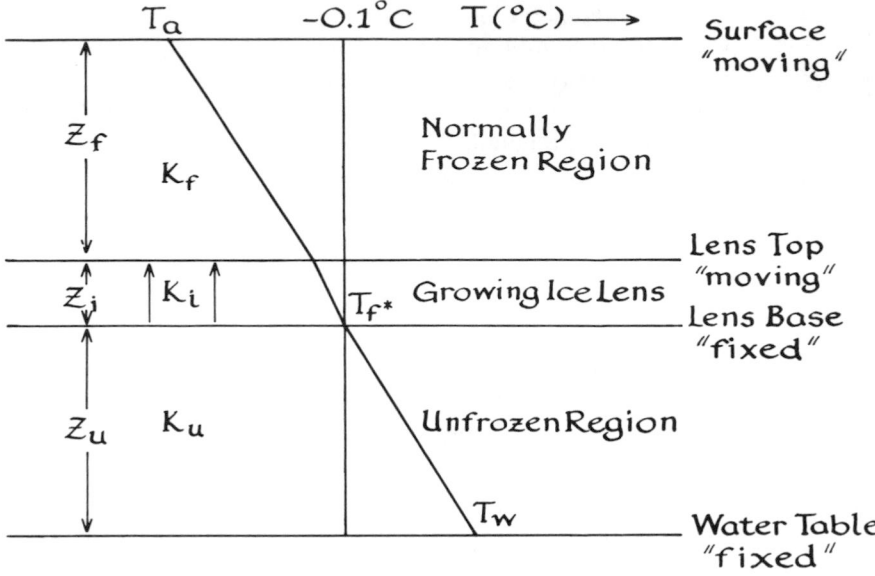

Fig. 7-2. Geometry during lens growth.

Q_u remain constant. It would therefore seem reasonable to calculate the heave rate as a function of the evolution of the heat flux system. To accomplish this it is necessary to calculate a thermal resistance (R) for each layer, assuming a thermal conductivity of 52×10^{-4} cal cm^{-1} sec $^{-1}$ °C^{-1} for the ice lens. This is accomplished by the following equations:

$R = Z/K$:	thermal resistance of a layer;	(6.1)
$Q_u = (T_w - T_{f*})/R_u$:	heat flux in the unfrozen zone;	(6.2)
$Q_f = (T_{f*} - T_a)/(R_f + R_i)$:	evolving heat flux in the normally frozen zone and ice lens;	(6.3)
$divQ = Q_f - Q_u$:	heat flux divergence at the lens base;	(6.4)
$L_f = \rho_i L + 0.5C_i(T_{f*} - T_a)$:	heat of fusion and undercooling, volumetric heat capacity of ice (C_i);	(6.5)
$\Delta Z_i = (divQ\Delta t)/L$:	heave increment (ice lens growth) during period t.	

As Q_u and R_f do not change during the period of growth, it is possible to write an iterative expression for the thickness of the ice lens. Here the index, n, refers to the time increment (Δt).

$$Z_i(n+1) = Z_i(n) + (\Delta t/L) \left[\frac{T_f - T_a}{R_f + Z_i(n)/K_i} - Q_u \right] \quad (7)$$

Note that, at the onset of heave, the initial heave due to normal freezing above the starting or reference surface is $((\rho_w/\rho_i) - \rho_w)Z_f V_p$. At any time during the continued growth of the lens the elevation, $H(n)$, above the reference surface can be expressed as a function of Z_u and Z_f (which remain constant during segregation), of the initial depth to the water table (Z_w), and of the current lens thickness, $Z_i(n)$:

$$H(n) = Z_u + Z_f + Z_i(n) - Z_w \quad (8)$$

TERMINATION OF LENS GROWTH

It can be seen from equation (7) that as the lens grows the incremental growth ($\Delta Z_i/\Delta t$) must decrease. This corresponds to heave records under stable environmental conditions. Note that in equation (7) the heat necessary to undercool the growing lens and normally frozen layer is included. Note further that when the lens has grown to a thickness which yields a flux divergence of zero, lens growth will stop. Equations (5) to (8) can therefore be employed in an iterative scheme to yield both an estimate of total heave and the duration of heave.

SIMULATION OF THE INFLUENCE OF VARIABLE WATER TABLE DEPTH AND
SURFACE TEMPERATURE ON HEAVE EVOLUTION

The model described above was used to study the time dependent behaviour of the ice segregation system using ten initial water table depths, from 10 cm to 100 cm in 10 cm increments, and ten surface temperature increments, from -2°C to -20°C in 2°C increments. This yielded 10 × 10 output matrices showing the depth of the normally frozen zone, the time until the beginning of ice lens growth, the final thickness of the ice lens and the length of time during which segregation growth continues. In these runs the initial time increment used in the second part of the model was 900 seconds. After the first 70 days this increment was increased by 5% at each iteration until a zero flux divergence condition was reached and lens growth stopped. Heave records for the first 70 days after segregation began are displayed in figure 7-3 for selected initial water table depths.

DISCUSSION OF MODEL RESULTS

The model output matrices appear in Table 7-1. At each water table depth, the depth of the normally frozen zone is increased at reduced surface

WATER TABLE = 10 cm.
Heave in Centimeters
Time in Days
80
70
60
50
40
30
20
10
20
30
40
50
60
20°c
18°
16°
14°
12°
10°
8°
6°
4°
2°c

WATER TABLE = 50 cm.
Heave in Centimeters
Time in Days
80
70
60
50
40
30
20
10
20
30
40
50
60
20°c
18°
16°
14°
12°
10°
8
6
4
2°

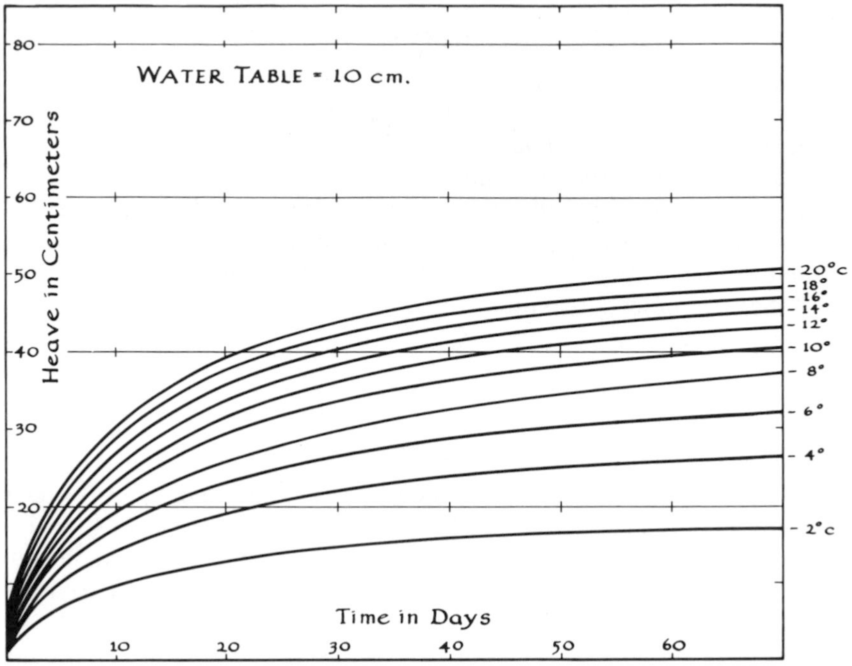

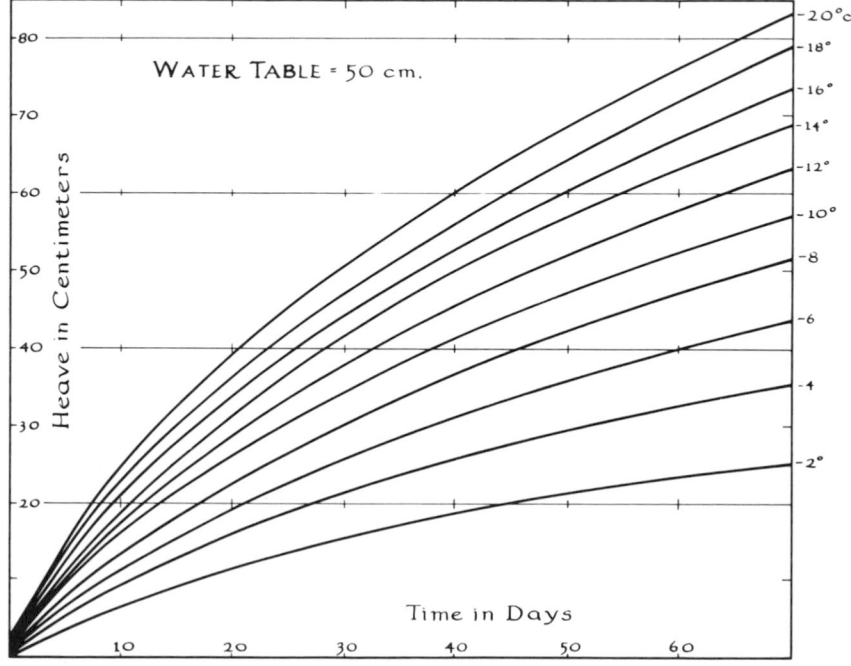

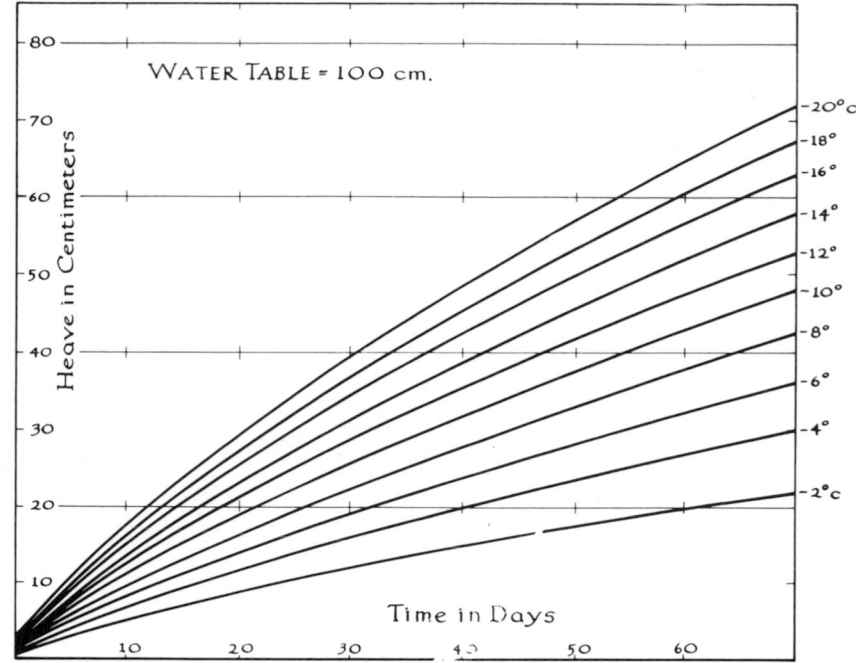

Fig. 7-3. Heave curves during the first 70 days of segregation ice formation for various water table depths.

temperatures. The duration of normal frost at each water table depth reaches a maximum at -6°C. This is a result of the fact that the magnitude of thè surface temperature disturbance appears both in the numerator and denominator of the Stefan Equation. The total heave as limited by zero flux divergence in the freezing zone increases with deeper water tables and reduced surface temperatures. The duration of segregation frost as limited by the zero flux divergence condition shows irregular local maxima at critical combinations of water table depths and surface temperature depressions. Recall, however, that the heave rate decreases rapidly with time.

Heave rates as a function of time are shown for water table depths of 10, 50 and 100 cm in figure 7-3. At the 10 cm water table depth, all the heave curves are approaching a near-zero time derivative condition after 70 days of ice segregation, although the zero flux divergence condition will not be reached for 1.89 to 3.91 years (see Table 7-1). With the water table reduced to 50 cm, the heave curves are nearly linear and strongly rising. The largest heave at the end of 70 days of ice lensing was produced by a surface temperature depression of -20°C at the 50 cm water table depth. The pattern at a water table depth of 100 cm is similar but heave rates are reduced. At water table depths greater than 50 cm the zero flux divergence condition is not reached for 25 to 922 years, with total heave between 89 and 512 cm.

TABLE 7-1

SIMULATION MATRICES

a. Depth of Normally Frozen Layer (cm)

w.t. depth (cm)	Surface Temperature (°C)									
	-2	-4	-6	-8	-10	-12	-14	-16	-18	-20
10	2.78	4.45	5.52	6.26	6.80	7.22	7.55	7.83	8.05	8.24
20	5.57	8.90	11.04	12.52	13.60	13.96	15.11	15.65	16.11	16.49
30	8.35	13.36	16.56	18.78	20.42	21.69	22.67	23.48	24.16	24.73
40	11.13	17.81	22.07	25.04	27.22	28.90	30.23	31.31	32.21	32.98
50	13.92	22.26	27.59	31.30	34.02	36.12	27.78	39.14	40.26	41.22
60	16.70	26.71	33.11	37.58	40.83	43.34	45.34	46.96	48.32	49.47
70	19.48	31.17	38.63	43.81	47.64	50.57	52.90	54.79	56.37	57.71
80	22.27	35.62	44.15	50.08	54.44	57.78	60.45	62.62	64.42	65.96
90	25.05	40.07	49.66	56.34	61.25	65.01	68.01	70.44	72.48	74.20
100	27.83	44.52	55.17	62.60	68.05	72.24	75.57	78.25	80.53	82.44

b. Duration of Normal Frost (days)

w.t. depth (cm)	Surface Temperature (°C)									
	-2	-4	-6	-8	-10	-12	-14	-16	-18	-20
10	0.10	0.13	0.14	0.13	0.13	0.12	0.12	0.11	0.11	0.10
20	0.41	0.53	0.55	0.54	0.51	0.46	0.47	0.44	0.42	0.41
30	0.91	1.19	1.23	1.21	1.16	1.11	1.05	1.00	0.95	0.91
40	1.62	2.11	2.19	2.15	2.06	1.96	1.87	1.78	1.69	1.62
50	2.53	3.29	3.42	3.35	3.22	3.07	2.91	2.78	2.65	2.53
60	3.65	4.74	4.93	4.84	4.63	4.41	4.20	4.00	3.81	3.65
70	4.97	6.45	6.71	6.57	6.31	6.01	5.72	5.44	5.19	4.96
80	6.49	8.43	8.77	8.58	8.24	7.84	7.46	7.11	6.78	6.48
90	8.21	10.67	11.09	10.87	10.42	9.93	9.45	8.99	8.58	8.21
100	10.14	18.17	13.69	13.41	12.87	12.26	11.67	11.10	10.59	10.13

TABLE 7-1

SIMULATION MATRICES

c. Total Heave Limited to Zero Flux Divergence (cm)

w.t. depth (cm)	Surface Temperature (°C)									
	-2	-4	-6	-8	-10	-12	-14	-16	-18	-20
10	17.72	28.33	35.08	39.73	43.11	45.63	47.70	49.10	50.27	51.20
20	35.44	56.65	70.15	79.45	86.34	99.67	95.21	98.20	100.52	102.33
30	53.16	84.96	105.23	119.19	129.39	136.65	142.74	147.25	150.80	153.47
40	70.88	113.31	140.31	158.92	172.42	182.56	190.33	196.33	201.09	204.67
50	88.60	141.64	175.38	198.62	215.55	228.20	238.10	245.51	251.38	255.74
60	106.33	169.96	210.45	238.04	258.61	273.88	285.54	294.63	301.66	306.94
70	124.03	198.28	245.54	278.18	301.73	319.47	333.13	343.66	351.93	358.08
80	141.76	226.62	280.62	317.83	344.88	365.42	380.77	392.83	402.16	409.17
90	159.48	254.96	315.71	357.51	387.92	410.80	428.20	441.93	452.37	460.23
100	177.20	283.27	350.87	397.24	431.02	456.35	475.78	491.52	502.66	511.57

d. Duration of Segregation Frost (years)

w.t. depth (cm)	Surface Temperature (°C)									
	-2	-4	-6	-8	-10	-12	-14	-16	-18	-20
10	1.89	3.91	1.98	1.98	1.98	1.98	1.98	3.73	2.07	3.91
20	5.17	5.96	6.55	6.55	5.68	6.25	5.42	4.71	4.49	4.29
30	27.67	18.79	13.41	12.78	12.78	12.18	11.06	11.06	10.05	9.58
40	54.59	57.31	29.04	29.04	21.72	20.69	18.79	17.90	17.06	16.25
50	54.59	57.31	57.31	57.31	57.31	57.31	30.48	29.04	26.36	25.11
60	60.16	60.16	60.16	60.16	921.76	113.28	465.65	57.31	57.31	33.59
70	63.16	80.56	84.58	921.76	224.09	193.60	235.28	76.73	235.28	60.16
80	84.58	102.76	184.39	151.73	465.65	921.76	131.10	315.23	107.89	102.76
90	107.89	131.10	235.28	443.49	465.65	137.65	921.76	235.28	131.10	330.99
100	124.87	443.49	443.49	184.39	224.09	151.73	235.28	315.23	235.28	131.10

ACKNOWLEDGEMENTS

This research was carried out as part of a joint research project between Pennsylvania State University and The University of Michigan entitled "Modeling physical and thermal disturbance in permafrost terrain of Northern Alaska". The work was sponsored by the Office of Polar Programs, U.S. National Science Foundation, Washington, D.C. (NSF-TPSU-UM-22206-131).

The idea of producing this simple solution is the result of discussions of ice lensing theory held at the U.S. Army Cold Regions Research and Engineering Laboratory during January 1977. This workshop was organized by Dr. Shunsuke Takagi of CRREL. The author is indebted to Dr. Gary Guymon of the Irvine Campus of the University of California for many helpful comments on this topic. The paper was presented at the second International Symposium on Ground Freezing, held at the Norwegian Institute of Technology, Trondheim, Norway, on June 24-26, 1980 (Preprints volume: 515-524).

REFERENCES

Arakawa, K. 1966. Theoretical studies of ice segregation in soils. J. Glaciology 6: 255-260.

Palmer, A.C. 1967. Ice lensing, thermal diffusion and water migration in freezing soils. J. Glaciology 6: 681-694.

Stefan, J. 1891. On the theory of ice formation in the Polar Sea (in German). Ann. Physik, 3rd ser., 4: 269-286.

Terzaghi, K. 1952. Permafrost. Boston Soc. Civil Engineers, J. 39: 319-368.

8
RECENT OBSERVATIONS ON THE DEFORMATION OF ICE AND ICE-RICH PERMAFROST

Norbert R. Morgenstern

Department of Civil Engineering, The University of Alberta
Edmonton, Alberta, T6G 2G7 Canada

ABSTRACT

Both ice and ice-rich permafrost behave in a time dependent manner when subjected to load. The relationship between strain rate and stress is known as a flow law. This flow law is of interest to geotechnical engineers for the evaluation of the long-term behaviour of foundations, slopes, and underground excavations in ice and ice-rich frozen ground. Unfortunately, few of the test data derived from glacier studies have been within the stress and temperature ranges relevant to most civil engineering problems and new experiments have been necessary. In addition it has been necessary to correlate laboratory findings with field performance data in order to evaluate the effectiveness of laboratory testing in establishing a flow law. This presentation summarizes recent laboratory and field investigations directed toward evaluating the flow law of ice and ice-rich permafrost at relatively warm temperatures.

RÉSUMÉ

Des observations récentes sur la déformation de la glace et du pergélisol riche en glace

La glace et le pergélisol riche en glace se comportent d'une manière qui dépend du temps quand ils sont soumis à une charge. Le rapport entre la vitesse de déformation et la force est défini comme loi d'écoulement. Cette loi intéresse les ingénieurs en géotechnique dans l'évaluation de la performance à longue durée des fondations, des pentes, et des excavations souterraines dans la glace et dans le pergélisol riche en glace. Malheureusement, peu des données expérimentales dérivées d'études des glaciers sont comprises dans les limites de tension et de température pertinentes à la plupart des problèmes en génie civile, ce qui a nécessité des expériences nouvelles. En plus, il importait de corréler les résultats de laboratoire avec les données de terrain pour évaluer l'efficacité des épreuves en laboratoire en vue d'établir une loi d'écoulement. Cet article résume les travaux récents en laboratoire et sur le terrain dont le bout est l'évaluation de la loi d'écoulement de la glace et du pergélisol riche en glace à des températures relativement élevées.

СОВРЕМЕННЫЕ НАБЛЮДЕНИЯ НАД ДЕФОРМАЦИЕЙ ЛЬДА И ЛЬДОНАСЫЩЕННОЙ МНОГОЛЕТНЕЙ МЕРЗЛОТЫ

Н. Р. МОРГЕНСТЕРН

РЕЗЮМЕ

Подвергнутые нагрузке лед и льдонасыщенная многолетняя мерзлота ведут себя времязависимым образом. Зависимость темп деформации от напряжения общеизвестна как закон потока. Этот закон потока интересует инженеров геотехников при оценке долговременного поведения фундаментов, склонов и подземных раскопок во льду и в льдонасыщенных мерзлых породах. К сожалению, очень мало данных, полученных из исследований ледников, находятся в диапазоне напряжений и температур, представляющих интерес с точки зрения большинства проблем строительства и поэтому нужно было организовать новые экспериментальные исследования. К тому же, надо было взаимоувязать результаты исследований в лаборатории с информацией полученной при полевых работах для того, чтобы оценить эффективность проверочных работ в лаборатории при определении закона потока. Настоящий доклад суммирует современные исследования в лаборатории и в полевых условиях, направленные на оценку закона потока когда его применяют ко льду и льдонасыщенной многолетней мерзлоте при сравнительно теплых температурах.

AN APPRECIATION

It is a distinct privilege to be invited to participate in this symposium honouring Ross Mackay. I hope that among the many accolades that Ross has received he will take it as a compliment if I describe him as an engineer's geographer. I say this because he has been relentlessly quantitative both in observing northern geomorphological processes and in explaining them. As one who must be quantitative in undertaking civil engineering works in the North and as someone who is sensitive to the need to understand the origins and distribution of ground ice in order to carry out rational design in permafrost, I will invariably turn to the writings of Ross Mackay to see what he has said about a particular ground ice form or genetic process. Ross Mackay has always given us both time and wise counsel in his characteristically generous manner. Both Northern Engineering and Northern Sciences are greatly in his debt.

INTRODUCTION

In our permafrost engineering work at the University of Alberta we have tended to perceive the subject as having a three-fold division, as follows:
1) Thawing ground, where we have been concerned with the development of thaw-consolidation theory and understanding the strength and deformation properties exhibited by ground subjected to thaw. Examples of applications include slope stability, stability of buried oil pipelines, and behaviour of oil well casings through permafrost.
2) Freezing ground, where we have been concerned with establishing an understanding of the frost heave process cast in a form suitable for engineering evaluations of the amount and rate of heave of ground subjected to freezing. Examples of applications include the prediction of heave of highways exposed to seasonal freezing and the behaviour of a chilled gas pipeline buried in unfrozen ground.
3) Frozen ground, where we have been concerned with the strength and deformability of frozen ground, particularly the more troublesome ice-rich soils found at relatively warm temperatures. These properties are needed to evaluate slope stability and to design foundations and underground excavations in frozen ground. In particular, as this paper indicates, we have been concerned with the flow law of ice-rich soil; this is a relation between stresses and strain rates.

Geotechnical engineering is concerned primarily with the design and execution of engineering works with and in natural materials. As such, the methodology adopted to facilitate a forecast of the behaviour of a geotechnical material when loaded goes beyond the simple concepts of selecting a specimen, testing it in the laboratory, describing its behaviour mathematically and then predicting how it might behave *in situ*.

An understanding of geotechnical behaviour comes, first, through experimenting with artificial materials in the laboratory. This maximizes control and reproductibility while avoiding the difficulties of sample disturbance and sample representation. However, only general insights into geotechnical properties can be obtained from artificial materials and the actual geotechnical substance, be it soil, rock or frozen ground, must be brought into the laboratory for evaluation. Even this has intrinsic limitations because laboratory specimens will usually be free of the non-homogeneity and anisotropy of the mass *in situ* but will still display disturbance effects resulting from sampling procedures. Minor geological details that exercise a dominant influence on mass behaviour are difficult to bring into the laboratory. Therefore, for the geotechnical engineer to develop confidence in his evaluation of material behaviour, laboratory studies must proceed in tandem with field studies. The geotechnical properties of any material are exposed to continual re-evaluation as a result of this procedure. In the case of some materials, such as soft clays, the process is quite mature and geotechnical properties can often be fixed with an accuracy sufficient for most practical purposes. However, in the case of naturally occurring frozen ground, the process is still in its infancy.

This paper, then, is more a progress report than a final statement. It synthesizes recent studies, conducted primarily at the University of Alberta, whose objective has been to determine the flow law of ice and ice-rich frozen ground in the stress and temperature range of greatest geotechnical interest.

BACKGROUND STUDIES

While experimental studies into the behaviour of frozen ground and ice have been underway in many countries for several decades, a convenient starting point here is the review paper presented by Anderson and Morgenstern (1973) to the Second International Permafrost Conference. It was pointed out that, with few exceptions, most of the test data on frozen soils dealt with reconstituted frozen sand, loaded at very low temperatures for relatively short durations. There was very limited information on the creep of natural permafrost and ice in the range of stress and temperature (i.e., close to 0°C) of practical interest. For ice, the classical studies of Glen (1955) were dominant. Only one set of data from field behaviour could be found (Thompson and Sayles, 1972) and this gave the paradoxical result that ice-rich Fairbanks Silt crept about three times faster than ice alone.

Figure 8-1 illustrates typical creep curves for frozen soil. Ice-poor soils display terminating creep at stress levels coincident with many engineering applications. While this behaviour merits study, its practical significance is less than the much greater deformations that develop with time in ice-rich soils. It seemed reasonable to assume that ice-rich soils might behave like ice and that one might be able to forecast the behaviour of frozen soils based on the data available in the literature describing the creep of ice. These data have been collected primarily for glaciological purposes.

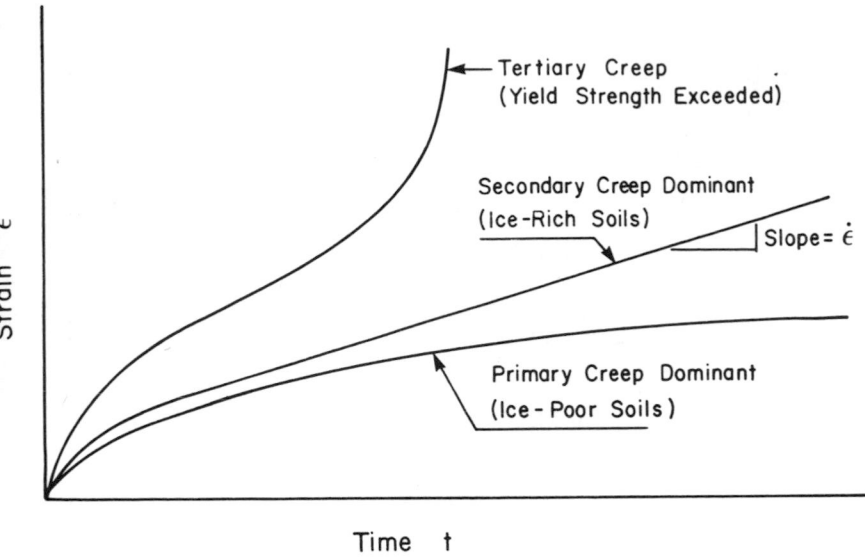

Fig. 8-1. Typical creep curves for frozen soil.

McRoberts (1973) undertook a comprehensive survey of the available data for ice relating the octahedral shear stress to the octahedral shear strain rate. The following relationship appeared to be the best fit:

$$\dot{\epsilon}_o = 0.04\tau_o^{1.1} + 0.08\tau_o^{4.5} \tag{1}$$

where ϵo denotes octahedral shear strain rate (1/year) and τ_o denotes octahedral shear stress (bars). A comparison of this relation with available data for secondary creep rates of frozen soils was generally inconclusive. In several instances it seemed that investigators had tended to overestimate secondary creep rates by stopping their tests prematurely. There was also a paucity of data at low shear stress levels.

McRoberts (1975) applied equation (1) to the analysis of an infinite slope. The results are reproduced in figure 8-2. If the creep law used in the analysis had merit, then surface velocities and their geomorphological expression would be as indicated in the figure. For example, a slope about 30 m high, inclined at 15° to the horizontal would be expected to display surface velocities in the order of 1 m a⁻¹. This did not seem to be in accord with our extensive experience with slope behaviour in ice-rich terrain within the

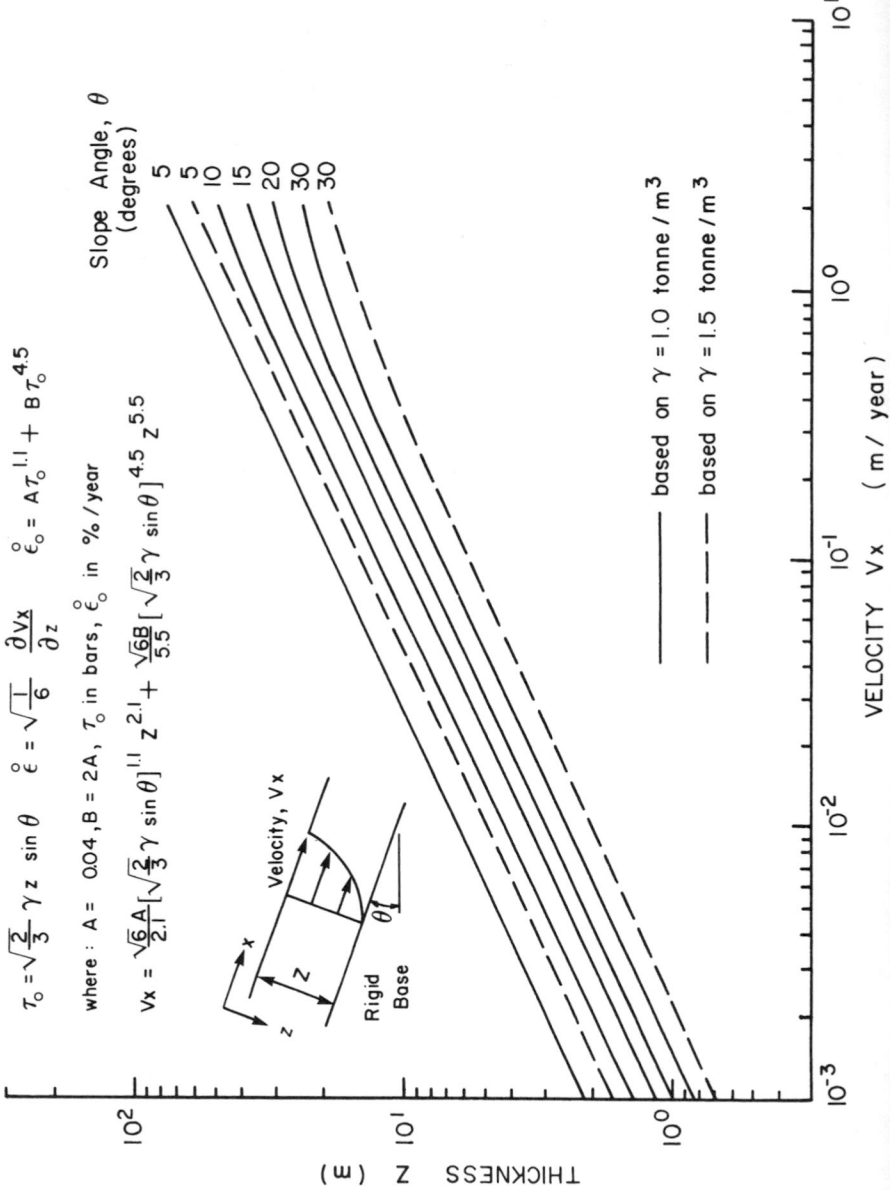

Fig. 8-2. Solution to the infinite slope using possible flow law for ice.

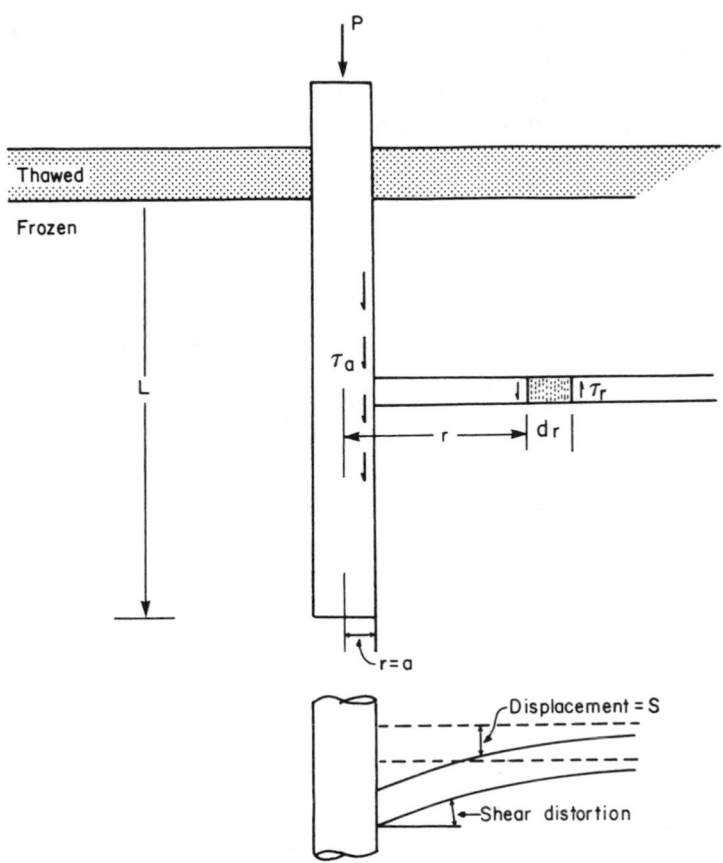

Fig. 8-3. Stresses and strains around a pile in permafrost.

Mackenzie River valley. It appeared that this synthesis of available data, and possibly the data themselves, did not provide a satisfactory basis for predicting the behaviour of naturally occurring ice-rich soils.

BEHAVIOUR OF A PILE IN ICE-RICH SOIL

Many foundations in permafrost are supported on piles and the long-term deformation of the foundation is a problem of considerable interest.

Nixon and McRoberts (1976) showed that the behaviour of a pile could be analyzed by treating it as a rigid rod in a non-linear viscous material. The mechanism is illustrated in figure 8-3.

The stress distribution for this mechanism can be obtained with relative ease and if the flow law of the frozen soil is known, it is possible to compare the predicted pile creep rate with the observed rate. Therefore, this analysis provided a means of designing piles on the basis of settlements and for investigating the validity of flow law data.

A new synthesis of the creep data for ice was undertaken. It was expressed in the same form as equation (1) but was extended to consider temperature dependence:

$$\dot{\epsilon}_o = A\tau_o{}^a + B\tau_o{}^4 \qquad (2)$$

where A, a, and B depend upon temperature. When applied to test data on pile creep, equation (2) had a tendency to over-predict, at least for the field cases where secondary creep behaviour had been established with some confidence. However, the design procedure was conservative, and an improvement over the current practice. Again, the limitation to the data base for the flow law became apparent.

At this time we became increasingly respectful of the difficulties associated with obtaining long-term creep data on naturally occurring ice-rich soils at temperatures close to freezing (Roggensack, 1977) and accordingly more critical of the interpretation of available literature on creep. For example, to adopt data as compatible with secondary creep behaviour, the creep rate must be constant with time. It is easy to convince oneself that this is so from an apparently linear portion of a plot of strain against time in a constant stress test. However, a more searching test is to plot strain rate against time as shown in figure 8-4. Only when the strain-rate becomes constant is secondary behaviour established in an unambiguous manner.

One of the key data sets in the literature, particularly for ice at low stresses, is that given by Mellor and Testa (1969). A re-evaluation of the data, illustrated in figure 8-5, showed that secondary behaviour had not been attained in all tests. This would lead to an over-estimate of creep rates in stress ranges of geotechical interest.

A new evaluation of the ice creep law was undertaken by Nixon (1978) and by Morgenstern, Roggensack and Weaver (1980). In the latter study, relatively new laboratory data from Barnes, Tabor and Walker (1971) and analyses of the creep of ice caps were thought to be particularly valuable. This led to the flow law illustrated in figure 8-6. As indicated, at a given stress level, particularly at high temperatures, anticipated strain rates were markedly less than those proposed earlier by Nixon and McRoberts (1976).

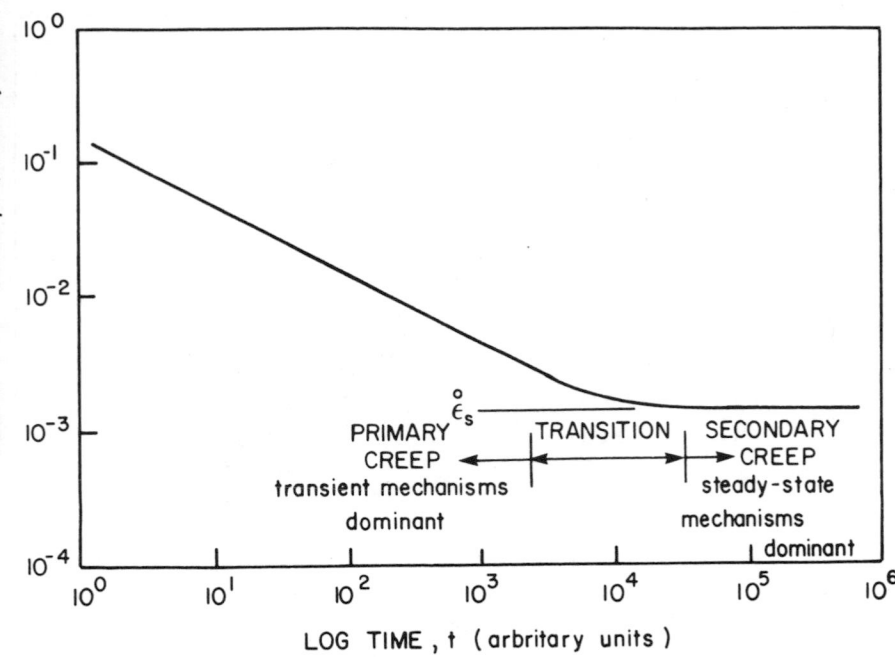

Fig. 8-4. Typical creep plot to evaluate transient and steady-state components of strain.

Available pile test data were also inspected carefully and only long-term test data were used in evaluating the applicability of the new flow law. The correspondence between predicted and observed behaviour is illustrated in figure 8-7. Given the variety of circumstances that lead to both the flow law and the pile test data, the agreement is remarkable. The flow law developed in this way is less restrictive in design than those listed earlier, and it appears consistent with the best available data. It is of the form

$$\dot{\epsilon}o = B\tau_o^3 \tag{3}$$

where B depends on temperature.

Students of glaciology will recognize immediately that equation (3) is no more than the equation proposed by Glen (1955). It appears that his equation provides an adequate representation of the secondary flow of ice and ice-rich soils, even at low stresses and relatively high temperatures. Insofar as the behaviour of ice constitutes an upper bound to that of ice-rich soil, use of equation (3) should lead to conservative results. However, fine-

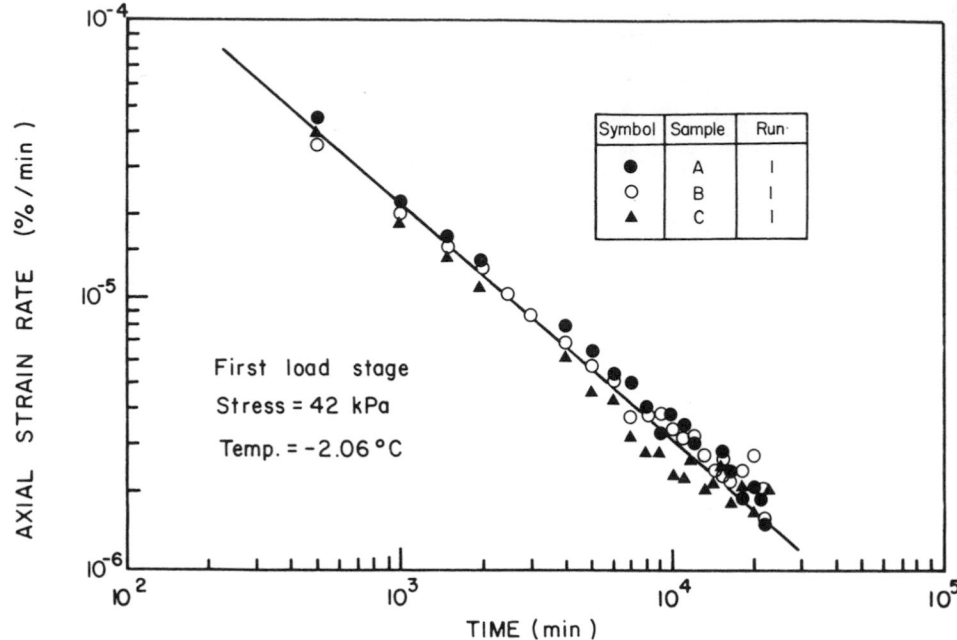

Fig. 8-5. Analysis of results from long-term creep tests. (data from Mellor and Testa, 1969).

grained soil close to melting possesses considerable unfrozen water and it is by no means certain that designing on the basis of ice will always be conservative. Moreover, further experimental confirmation of equation (3) at low stresses and for multi-axial stress conditions is still needed, and a systematic assessment of its limitations when applied to more field problems is mandatory. As noted earlier, it is only by means of comparison between field and laboratory behaviour that the geotechnical engineer gains confidence in his analytical skills.

NEW EXPERIMENTAL STUDIES ON ICE

New experimental studies into the creep behaviour of ice under low stresses have been conducted by Sego (1980). This test series involved carefully controlled, reproducible ice specimens tested under conditions of either constant stress or constant displacement rate. Multi-axial behaviour was investigated by performing direct simple shear tests and punch tests.

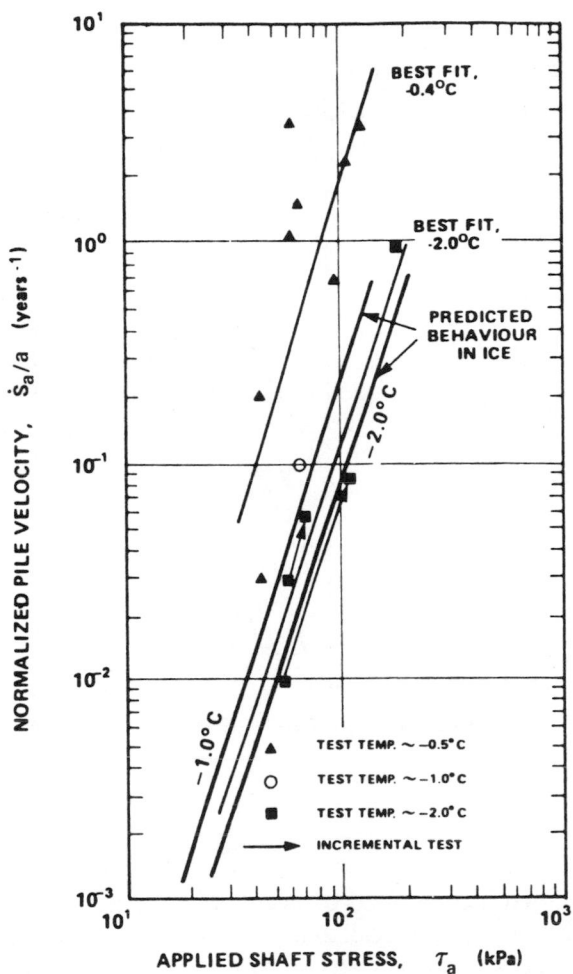

Fig. 8-6. Proposed flow law for ice.

The latter were predicted from the observed flow law for purposes of comparison.

Figure 8-8 presents typical data from a constant displacement rate experiment. Sego emphasized the significance of the observation that ice weak-

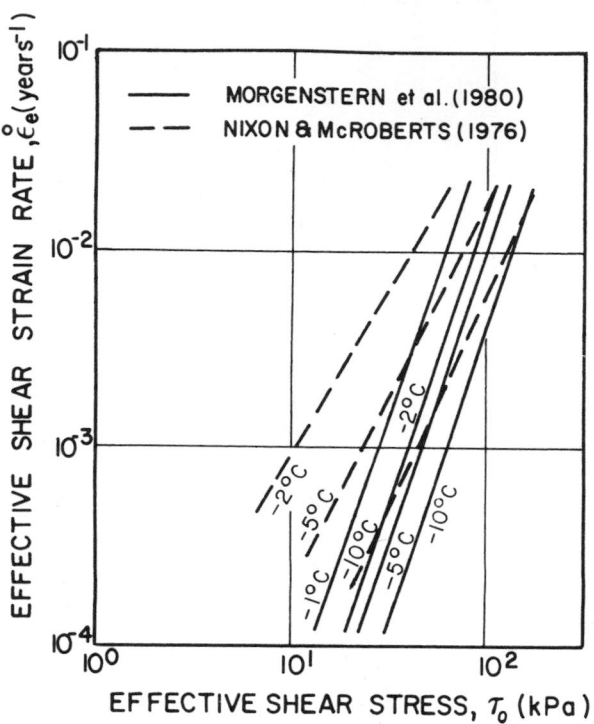

Fig. 8-7. Summary of long-term pile creep in ice and ice-rich soils.

ens under continued applied load or accumulated strain. In general, the resistance of ice drops from a peak at about 1% strain to an approximately constant resistance beyond 10% strain. This is analogous to the decrease from peak to residual resistance exhibited by clays. In the case of clays, this change in resistance is associated with particle orientation, while in the case of ice recrystallization and preferred crystal orientation occur at large strain.

Sego found that the flow at low strain differs from that at large (10%) strain. This is another important finding with engineering implications. Both take the form of equation (3), but at small strains the exponent is about 2.5. At large strains these new data correspond well with earlier work to confirm the use of the exponent 3. Figure 8-9 provides the comparison

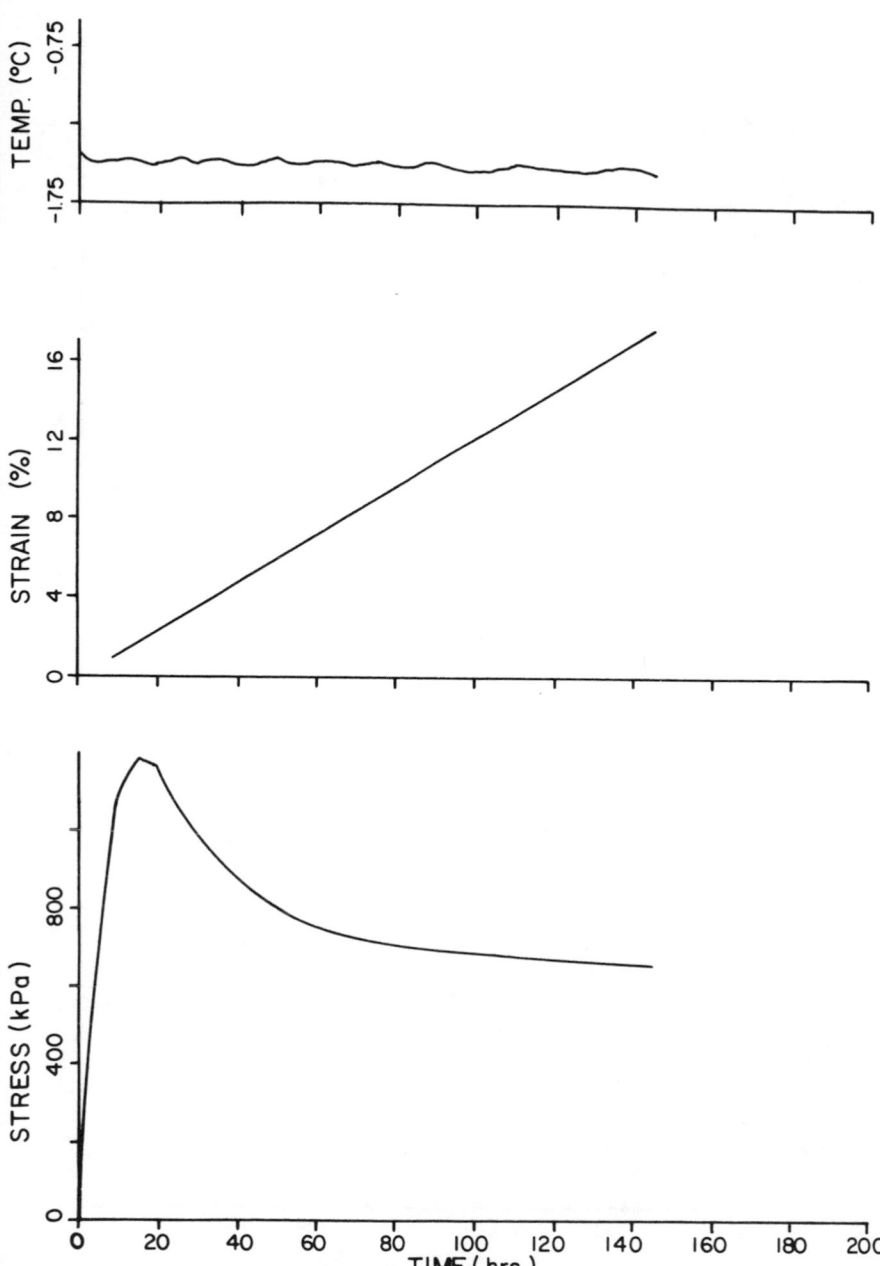

Fig. 8-8. Typical plot of data from the constant displacement rate experiment. (Sego, 1980).

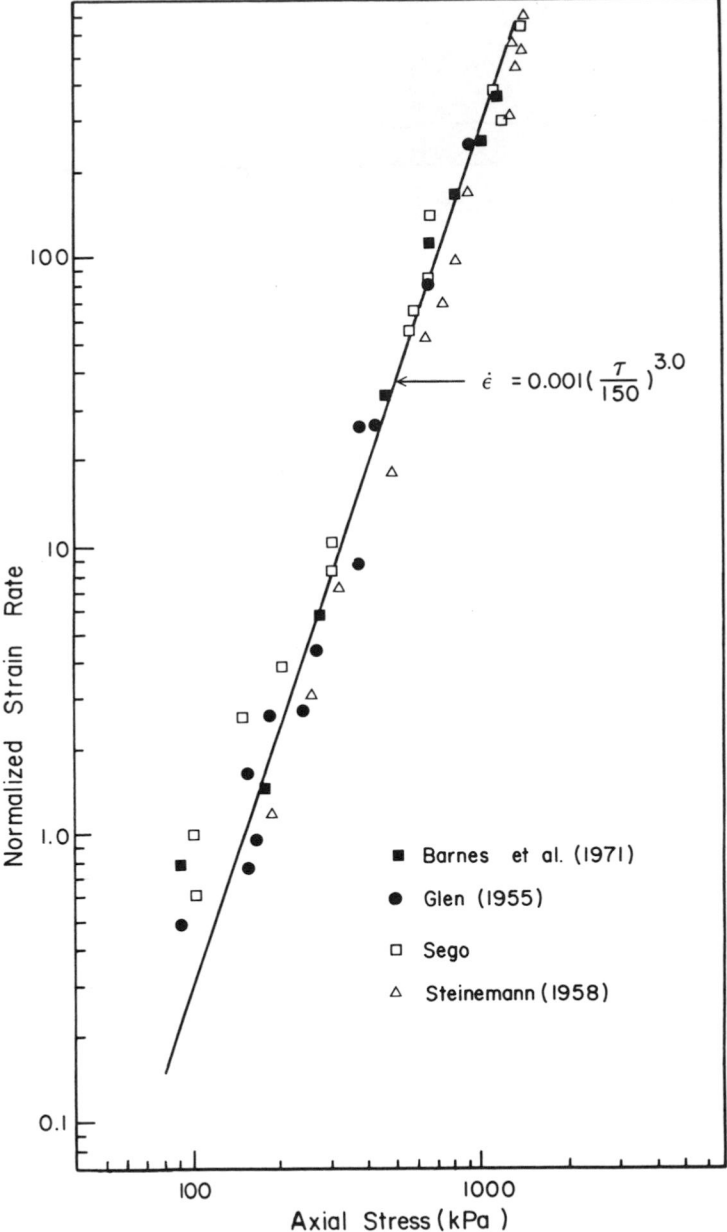

Fig. 8-9. Strain rate vs. stress normalized from grain size effects. (Sego, 1980).

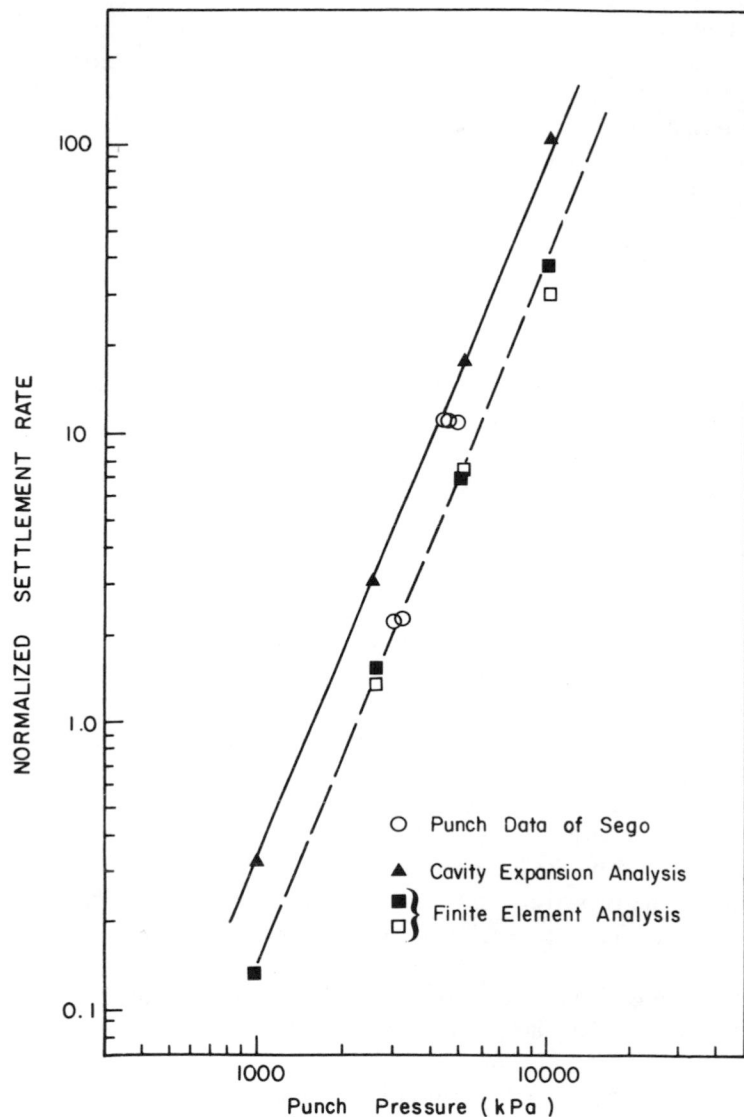

Fig. 8-10. Comparison between theoretical and laboratory data for punch tests on ice. (Sego, 1980).

between various investigations corrected for grain size effects, another finding from this study.

It is possible to extend to multi-axial conditions the results of uniaxial tests and thereby obtain a general relation between stresses and strain rates. Whether or not this generalization is correct remains a further issue for experimental verification. Sego performed constant velocity punch indentation tests at various pressures and compared the results with predictions based on such a generalization. A comparison of the data with alternative theoretical models is shown in figure 8-10. The correspondence is good. More details are available in Sego (1980).

As a result of this experimental work, our confidence in using a power law with an exponent of three for large deformation problems has been enhanced. Many other aspects, such as different small strain behaviour and grain size effects, have been drawn to our attention. However, for problems of continued creep, the strains, stresses and deformation calculated with such a power law (e.g., Nixon, 1978) appear to be a reasonable approximation suitable for engineering design.

NEW FIELD STUDIES ON ICE-RICH PERMAFROST

As noted earlier, the geotechnical engineer has an on-going interest in comparing the results of laboratory tests with behaviour in the field. In the present context, we are interested in evaluating how well equation (3) describes the *in situ* creep of ice-rich permafrost. Studies undertaken by Savigny (1980) and summarized by Morgenstern (1981) contribute to resolving this issue.

Savigny undertook a field investigation of creep in a natural slope composed in large part of relatively warm, fine-grained, ice-rich, structurally non-homogeneous permafrost soil. The site selected for instrumentation is on the south bank of Great Bear River, a major tributary of Mackenzie River. The site is about 7 km upstream from Fort Norman at the confluence of the two rivers and lies within the widespread discontinuous permafrost zone. The site was selected for several reasons:

i)   It was an intended crossing for a proposed major pipeline.
ii) It was among the highest and steepest slopes in fine-grained soils encountered in the Mackenzie Valley.
iii) The stratigraphy was characteristic of that of extensive areas of the Mackenzie Plain.
     The field studies had four main objectives:
i)   the installation of borehole inclinometers to measure *in situ* creep deformation in the ice-rich soils comprising the slope;
ii) the installation of thermistor strings to establish the temperature gradient affecting each inclinometer casing;
iii) the installation of piezometers below the base of the permafrost to

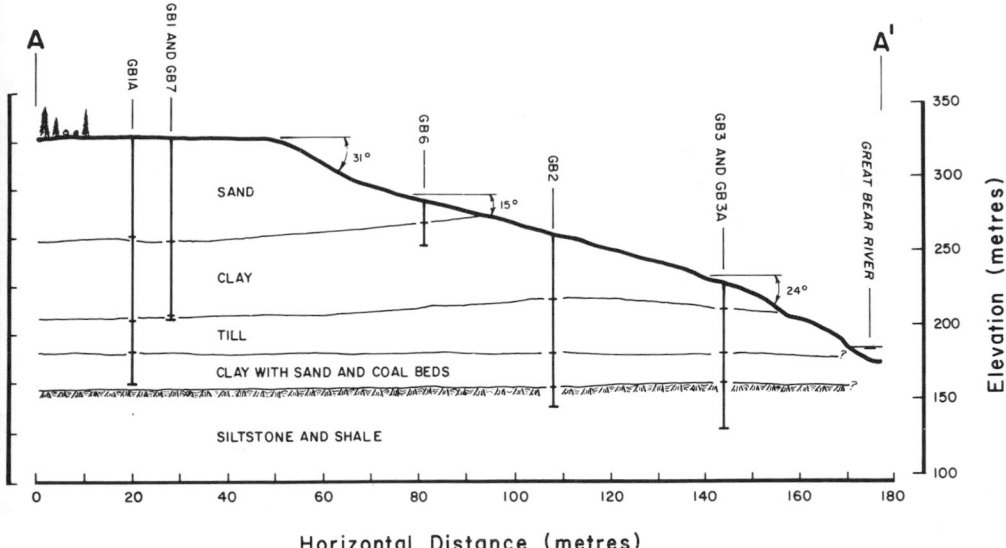

Fig. 8-11. Cross section of instrumented slope, left bank of Great Bear River, N.W.T., near the confluence with Mackenzie River.

assess the overall stability of the slope against deep-seated failure;
iv) to obtain continuous undisturbed cores from each hole in order to establish the stratigraphy, to determine basic soil properties and to permit detailed laboratory investigation of deformation properties under simulated field conditions.

Each of these objectives was met in varying degree. Detailed information is given by Savigny (1980).

Figure 8-11 presents a cross-section of the slope. Glaciolacustrine and glaciodeltaic sand overlie the till deposited by the Wisconsin Laurentide ice sheet. The till rests on an established clay, sand and coal deposit that appears to be alluvial in origin and lies unconformably over the Tertiary siltstone and shale bedrock. The clays are of great interest. They are dark grey, rhythmically laminated, medium to highly plastic, silty clay. They are fissured throughout and commonly slickensided in association with ice veins. Reticulate ice (Mackay, 1974) is the most common ice form but other more tabular forms are also present.

As a result of temperature observations, the thermal cross-section shown in figure 8-12 has been constructed. The data represent mean annual temperatures below the depth of zero mean annual temperature fluctuation. Piezometric observations indicate water levels at the base of the

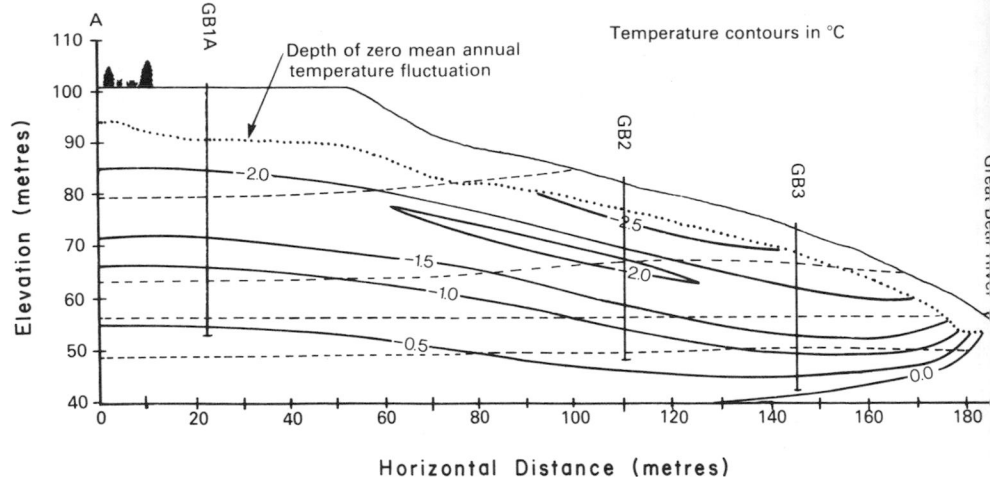

Fig. 8-12. Thermal cross-section of instrumented slope, Great Bear River.

permafrost controlled by the Great Bear River. Presumably, sandy zones, joints and thin sandstone laminae in the bedrock provide a means of rapid pore water communication.

In order to discern consistent downslope creep movements, both the installation and subsequent readings had to be made with the utmost care. Although it transpired that the instruments were being utilized at, or even beyond, their intended resolution, systematic downslope velocities in the range of 1-2 mm a^{-1} were detected. These are shown for one of the test holes in figure 8-13.

Using currently available methods of stress and deformation analysis it is possible to take equation (3) in a form reflecting average temperature conditions and predict the theoretical velocity profile. This prediction is also shown in figure 8-13 and it is seen that the calculated velocities far exceed those observed. If the modulus in the equation (B) is reduced by a factor of six and the calculations repeated, an acceptable correspondence between theory and observations is obtained. Therefore, we can conclude from this singular case history that at least for the glaciolacustrine clays common in the Mackenzie Valley, and at the relatively warm temperatures of engineering interest,

$$B \text{ (ice)} \approx 6B \text{ (ice-rich, fine-grained permafrost soil)} \qquad (4)$$

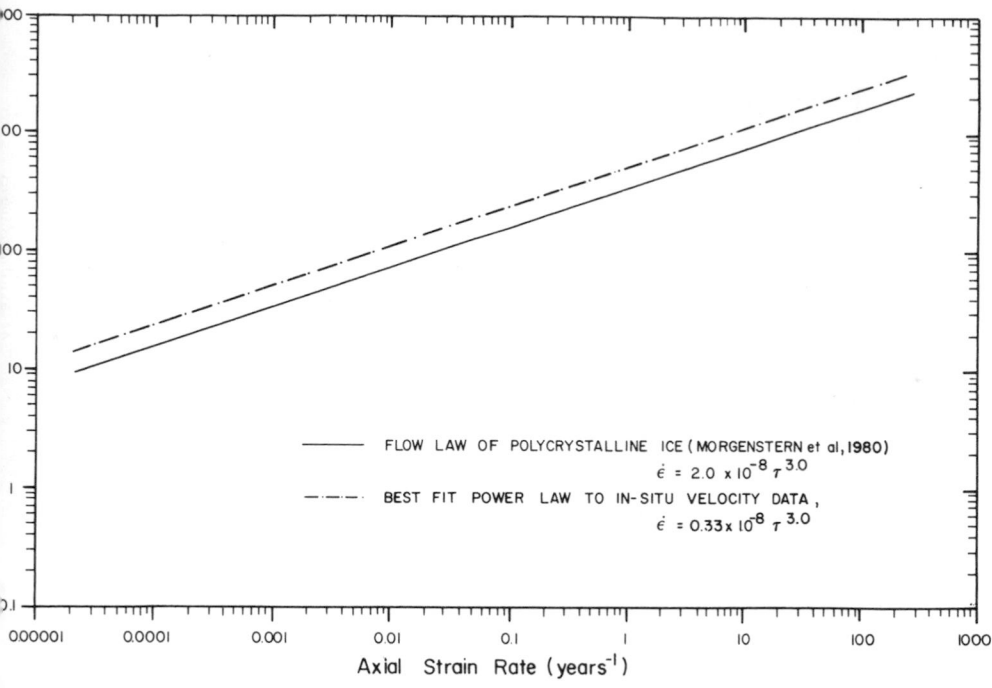

Fig. 8-13. Comparison of creep relations used in numerical analysis.

So far, the same flow law and exponent seem applicable for engineering purposes. A comparison between the flow law for ice and that deduced from back-analysis of the creep observations is given in figure 8-14.

CONCLUDING REMARKS

This presentation has summarized recent work undertaken to evaluate the flow law for ice and for ice-rich permafrost. For large strain problems early experiments on ice performed by Glen (1955) have been corroborated over a wide range of stress and temperature conditions. It appears that a power law with an exponent of three may be used in engineering calculations to assess long-term deformations. In the case of permafrost soils, a correction can be applied to the B term to account for the field observations that ice-rich permafrost is less disposed to secondary creep than ice.

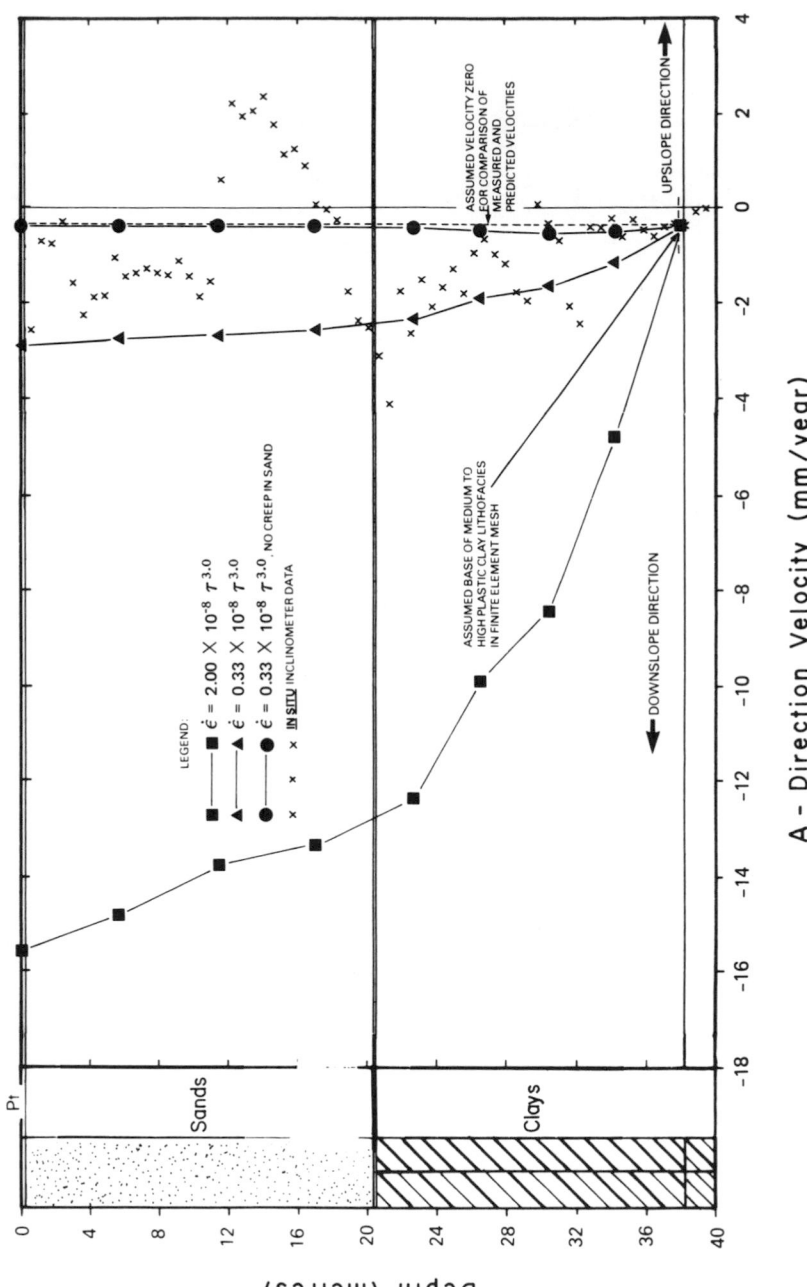

Fig. 8-14. Comparison of creep relationship derived from ice tests and back-analysis of slope creep.

However, many more studies on natural permafrost will be needed before the flow law can be stated with authority.

Ross Mackay's greatest achievements have been in the field. It is through quantitative observations and analysis of field phenomena that he has elucidated a variety of periglacial processes. The geotechnical engineer must emulate this if engineering is to be conducted with confidence in permafrost terrain.

REFERENCES

Anderson, D.M. and Morgenstern, N.R. 1973. Physics, chemistry and mechanics of frozen ground: a review. Permafrost 2nd Int. Conf., Yakutsk, U.S.S.R., July 13-28: North American Contribution. Washington, D.C., NAS (U.S.A.): 257-288.

Barnes, P., Tabor, D., and Walker, J.C.F. 1971. The friction and creep of polycrystalline ice. Roy. Soc. London, Proc. *324A*: 127-155.

Glen, J.W. 1955. The creep of polycrystalline ice. Roy. Soc. London, Proc. *228A*: 519-538.

Mackay, J.R. 1974. Reticulate ice veins in permafrost, northern Canada. Can. Geotech. J. *11*: 230-237.

McRoberts, E.C. 1973. Stability of slopes in permafrost. The University of Alberta, Edmonton. Ph.D. Thesis: 370 pp.

McRoberts, E.C. 1975. Some aspects of a simple secondary creep model for deformations in permafrost slopes. Can. Geotech. J. *12*: 98-105.

Mellor, M. and Testa, R. 1969. Creep of ice under low stress. J. Glaciology, *8*: 147-152.

Morgenstern, N.R. 1981. Geotechnical engineering and frontier resource development. Geotechnique *31*: 305-365.

Morgenstern, N.R., Roggensack, W.D. and Weaver, J.S. 1980. The behaviour of friction piles in ice and ice-rich soils. Can. Geotech. J. *17*: 405-415.

Nixon, J.F. 1978. First Canadian geotechnical colloquium: foundation design approaches in permafrost areas. Can. Geotech. J. *15*: 96-112.

Nixon, J.F. and McRoberts, E.C. 1976. A design approach to pile foundations in permafrost. Can. Geotech. J. *13*: 40-57.

Roggensack, W.D., 1977. Geotechnical properties of finegrained permafrost soils. The University of Alberta, Edmonton. Ph.D. Thesis: 449 pp.

Savigny, K.W. 1980. In-situ analysis of naturally occurring creep in ice-rich permafrost soil. The University of Alberta, Edmonton. Ph.D. Thesis: 439 pp.

Sego, D.C. 1980. Deformation of ice under low stresses. The University of Alberta, Edmonton. Ph.D. Thesis: 450 pp.

Thompson, E.G. and Sayles, F.H. 1972. In-situ creep analysis of a room in frozen soil. Amer. Soc. Civil Engineers, J. Soil Mechanics and Foundations Div. *98*: 899-915.

9
DISTRIBUTION OF RECENTLY ACTIVE ICE AND SOIL WEDGES IN THE USSR

N.N. Romanovskij

Faculty of Geological Sciences, Moscow State University
Moscow, 117234, U.S.S.R.

ABSTRACT

On the basis of field research and synthesis of prior observations, a map is constructed here of the distribution of recently active ice and soil wedges in the USSR. Conditions for the development of ice and soil wedges are reviewed in order to specify the principles which guide definitions and generalization in the map. Earth materials and climate fluctuations contribute substantial complexity to the observed pattern. Regional variations of wedge activity are reviewed and attention is drawn to the "ice complex" — the superimposition of various ground ice types, including thick, syngenetic ice wedges — which occurs in suitable soils in areas of extreme periglacial climate.

RÉSUMÉ
Distribution des coins de glace et des fentes de gel récemment actifs en URSS

Sur les bases de recherche sur le terrain et d'une synthèse d'observations antérieures, on a dressé une carte de localisation des coins de glace et des fentes de gel récemment actifs en URSS. Les conditions de développement des coins de glace et des fentes de gel sont revues afin de préciser les principes à la base des définitions et des généralisations sur la carte. Les matériaux et les variations climatiques contribuent de façon substantielle à la complexité de la distribution observée. Les variations régionales dans l'activité des fentes sont passées en revue et l'on attire l'attention sur 'le complexe de glace' — la surimposition de types variées de glace de sol, y compris les épais coins de glace syngénétiques — qui se forment dans des sols appropriés dans les zones à climat périglaciaire extrême.

РАСПРЕДЕЛЕНИЕ ПОСЛЕЛЕДНИКОВЫХ ДЕЯТЕЛЬНЫХ ЛЬДОВ И КЛИНЬЕВ ПОРОДЫ В СССР

Н. Н. РОМАНОВСКИЙ

РЕЗЮМЕ

На основании полевых исследований и синтеза данных из прошлых исследований, изготовлена карта распределения послеледниковых деятельных льдов и клиньев породы в СССР. Рассматриваются условия развития льдов и клиньев породы, с целью указать на закономерности определяющие исходные положения и обобщения приняты при изготовлении карты. Причем, состав пород и климатические перемены значительно усложняют установленное распределение. Анализируются региональные отклонения в поведении клиньев и обращается внимание на ''ледяные комплексы'' — накладывание разных типов погребенных льдов один на другой — включая мощные сингенетические ледяные клинья, встречаемые в соответствующих породах в районах подвергающихся действию крайне сурового перигляциального климата.

INTRODUCTION

Studies in the distribution of recently active ice and soil wedges are aimed at the following goals: first, conditions can be defined under which these wedges develop. These conditions include, above all, rock composition, moisture, and temperature regime. Second, the distribution of the process of frost cracking can be evaluated. This process governs the development of ice and soil wedges. Third, geological engineering problems can be attacked. Surfaces with active wedges are unfavourable for many kinds of construction and require specific approaches.

I deal here with the distribution of only two, genetically closely inter-related, types of wedge. Results of investigations are presented on a reference map (figure 9-1) showing the distribution of recently active ice and soil wedges. The map is based on personal research in various regions and a generalisation of the voluminous literature on the matter.

CONDITIONS FOR THE DEVELOPMENT OF WEDGES

To specify the subject of discussion, I present briefly the essence of ice and soil wedging and define the principles of map preparation and generalisation of facts. Ice and soil wedges are generated by frost cracking due to thermal contraction of frozen ground in winter. Frost cracks are

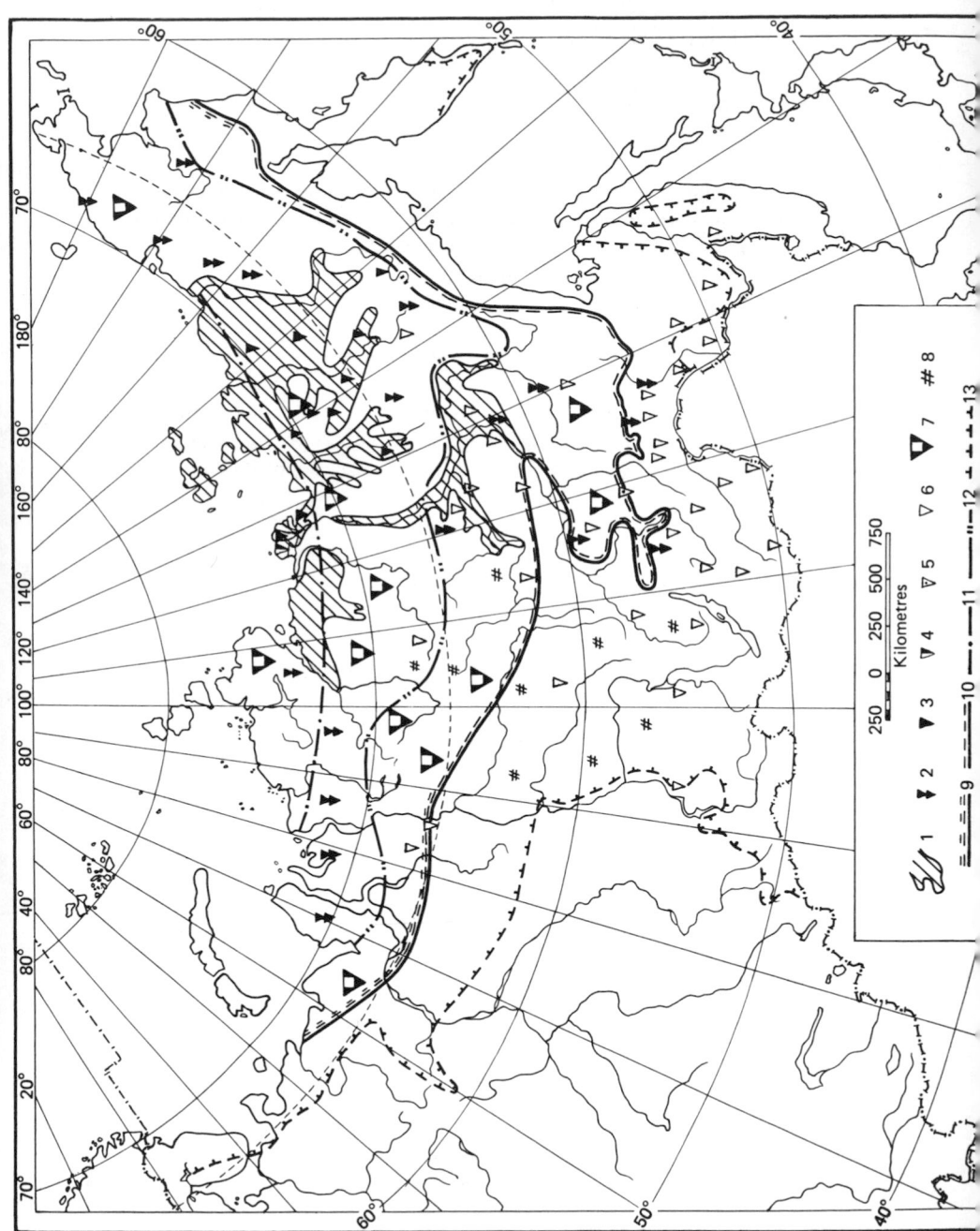

mainly filled in spring with snowmelt and river waters. This kind of infilling is most typical of northern Eurasia and North America. Other kinds of infilling, such as dry sand and weathered loam slumping from the walls of frost cracks, are of insignificant distribution under the conditions present now in these regions. Therefore, we do not specifically discuss the associated sand wedges and composite wedges, although areas of their present-day distribution are shown on the map.

Annual ice veins forming in frost cracks have different fates, depending on whether they stay in the active layer or penetrate into the permafrost. In the active layer, they thaw in summer to be replaced by a sediment that differs in composition, colour, density, and bedding from the surrounding material. In the permafrost, annual ice veins survive and ice wedges are formed in several years. The ice portion of a wedge is overlaid by its soil portion, whose thickness depends on the thickness of the seasonally thawed layer. Cracks restricted to the active (seasonally frozen and seasonally thawed) layer give rise to primordial soil wedges (figure 9-2).

Primordial soil wedges in rocks of one and the same composition generally form when the mean annual ground temperature at the level of zero annual amplitude, both positive and negative, approaches 0°C. The most important condition is a wide temperature range at the ground surface. Therefore, soil wedges are characteristic of continental areas both near the southern boundary of the permafrost area and beyond its limits.

Northwards in the permafrost area, as the temperature of the ground decreases, frost cracks begin to penetrate from the seasonally thawed layer into the permafrost. Annual ice veins survive in them. As a result, soil wedges are transformed into ice wedges. The decrease of ground temperature promotes a deeper penetration of cracks below the base of the

Fig. 9-1. Reference map showing the distribution of recently active ice wedges and primordial soil wedges in the U.S.S.R. Key: . . . .
1. area of widespread ice complex with thick, syngenetic ice wedges;
2. distribution of ice complex in river valleys;
3. recently active ice wedges;
4. recently active wedges with cracks filled with sand and ice in permafrost (low temperature composite wedges);
5. recently active wedges in the active layer with cracks filled with sand (high-temperature wedges);
6. recently active, primordial soil wedges;
7. frost polygons on residual soils with frost sorting of stones;
8. small diameter polygons (1.5 - 3.0 m);
9. boundary of areas of active ice wedges in peat bogs;
10. boundary of areas of active ice wedges in peat-bearing sandy loam and loam;
11. boundary of areas of active ice wedges in coarse-grained sand, gravel and pebbles;
12. southern boundary of frost polygons;
13. southern boundary of permafrost.

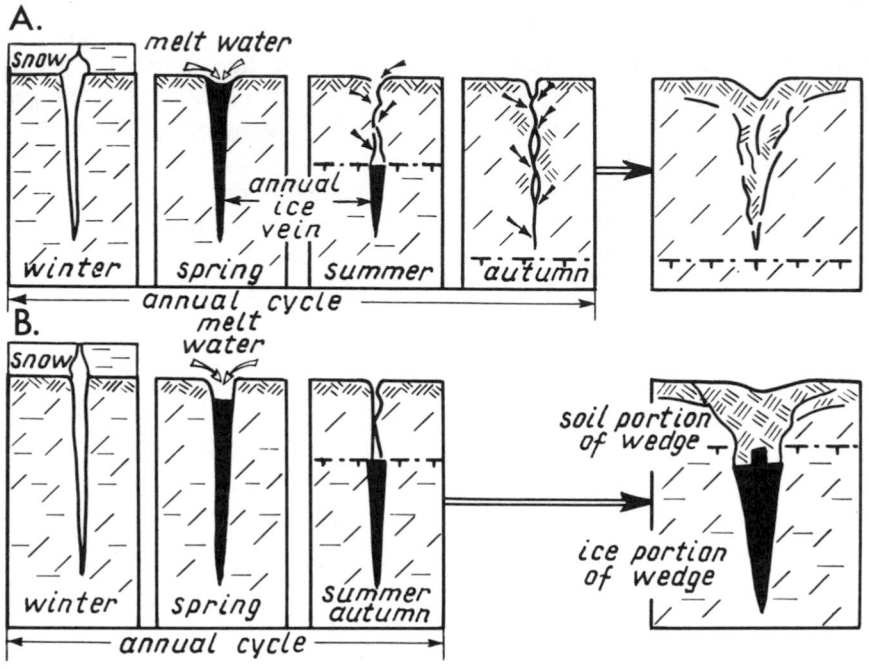

Fig. 9-2. The scheme of soil wedge (A) and ice wedge (B) formation.

seasonally thawed layer and an increase in the size of ice wedges. Consequently, permafrost zonation is observed in the distribution of soil and ice wedges, which is precisely shown on the map.

Significantly, this zonation is easily observed only in earth materials of similar composition and ice content; in other words, in similar facies and landscapes, namely, on marshy floodplains underlain by loam, on dry surfaces of alluvial, aeolian, and glaciofluvial sands, or on peat bogs. In rocks of different compositions, the transition from primordial soil wedges to ice wedges takes place within different temperature ranges. Moreover, ground temperatures may be quite different within the same region. Generally, fine grained sediments are cooler than sands and gravels. Hence in nature, the boundaries between the areas of wedges in peat bogs, loams, sands, and gravels do not coincide at all, and distances between them are occasionally hundreds of kilometres (figure 9-3). This has also been reflected on the map presented here, which shows two southern boundaries of areas of ice wedges; in loams and peat bogs on the one hand, and in sands and gravels on the other.

The boundaries as such are fairly diffuse; rather they are zones of transition of one type of wedge into the other. This is because wedges grow during a geologically prolonged time interval when the temperature of the ground and the depth of their seasonal thawing change many times. These variations occur within a definite, relatively narrow range. When the temperature decreases, the boundary shifts to the north, and when it increases, the boundary shifts to the south and a transition belt forms.

There are many specific features of soil and ice wedges connected with the fluctuation of the depth of the seasonally thawed layer (STL) or active layer in the transition belt. These features are shown in figure 9-4. Most obviously, voids form as a result of increasing STL and melting ice wedges during the period when the boundary shifts to the north. When the boundary shifts to the south, these voids are filled with thermokarst cave ice.

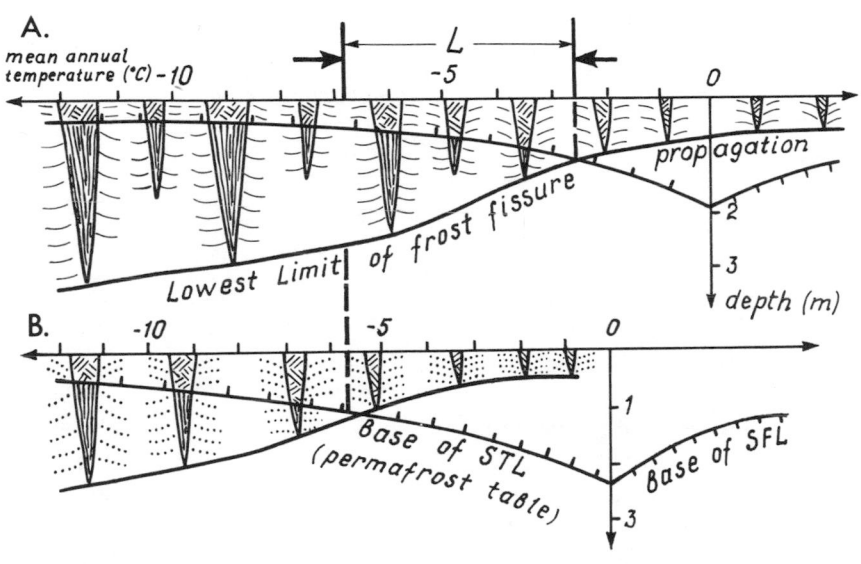

Fig. 9-3. The scheme of penetration of frost fissures and correlation between primordial soil wedges and ice wedges depending on soil mean annual temperature in loams (A), sands and gravels (B). L is the distance between boundaries of ice wedges in loam and of ice wedges in sands and gravels.

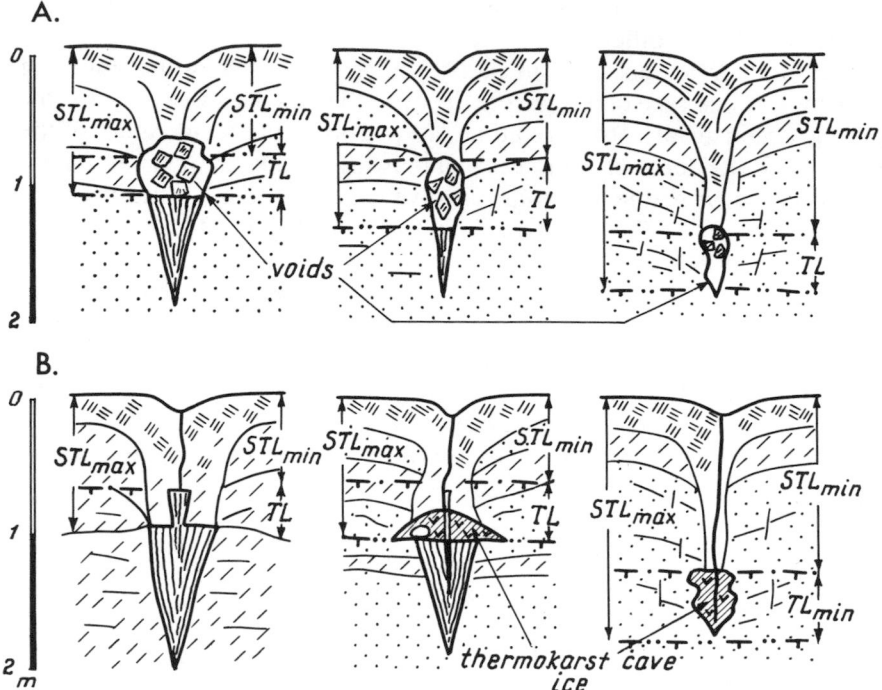

Fig. 9-4. Wedges of the transitional belt: (A) when the boundary shifts to the north, and (B) when the boundary shifts to the south.
STL - seasonally thawed layer; TL - transitional layer
$(TL = STL_{max} - STL_{min})$.

REGIONAL VARIATION OF WEDGE ACTIVITY

When generalising the materials that make up the basis for the map, the solution of two problems was essential: first, what is meant by recently active wedges and, second, how active wedges can be distinguished from inactive wedges. The solution to these problems was of special importance when we used the data obtained from publications by different authors in different years. In my opinion, recently active wedges form both in present-day sediments and in old, heterogenous ones, and especially in residual soils of widely varying age. Present-day sediments give rise to syngenetic, and old ones to epigenetic wedges. Ice wedges are regarded as active if they lie just below the seasonally thawed layer or exhibit direct manifestations of

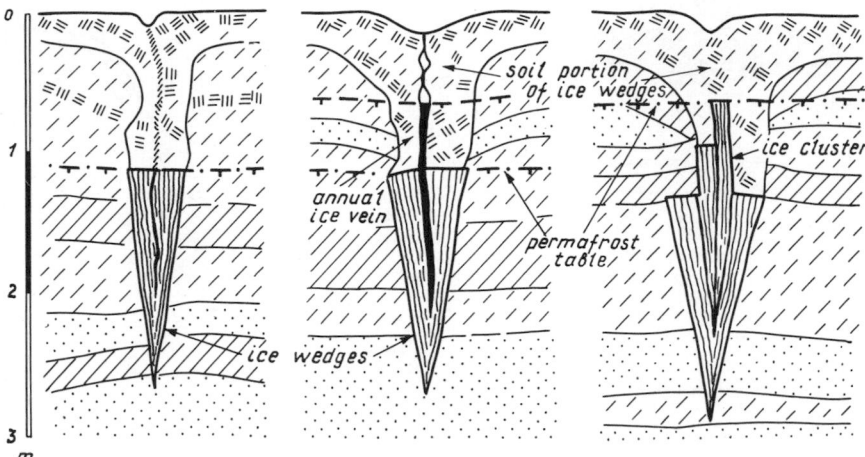

Fig. 9-5. Evidence of recent ice wedge activity.

their growth, namely, annual ice veins extending through the seasonally thawed layer. These veins and their clusters above the wedges prove that the wedges are now geologically active (figure 9-5). Both active and inactive wedges occur in polygonal systems of ice wedges. However, even if a small fraction of wedges provides evidence for their growth, we consider the system active.

The activity of soil wedges is more difficult to determine. Only direct observation of cracks and seasonal ice veins can provide reliable evidence for the recent development of soil wedges. But few such observations have been made to date. They are frequently conducted in metropolitan areas where snow is removed in winter and frost cracking is a result of man's activity — which occurs nowhere in nature. For instance, such cracks form on the square in front of the Moscow State University building in Lenin Hills. We consider a soil wedge recently active if its height is less than the depth of the active layer and it is seen in the relief. Because of these circumstances, the area of active soil wedges is shown on the map with a lower accuracy than the area of active ice wedges.

It is important to point to some regional features of the distribution of active ice and soil wedges. In the north of the European USSR and in West Siberia, active ice wedges are restricted, at the southern limit of their

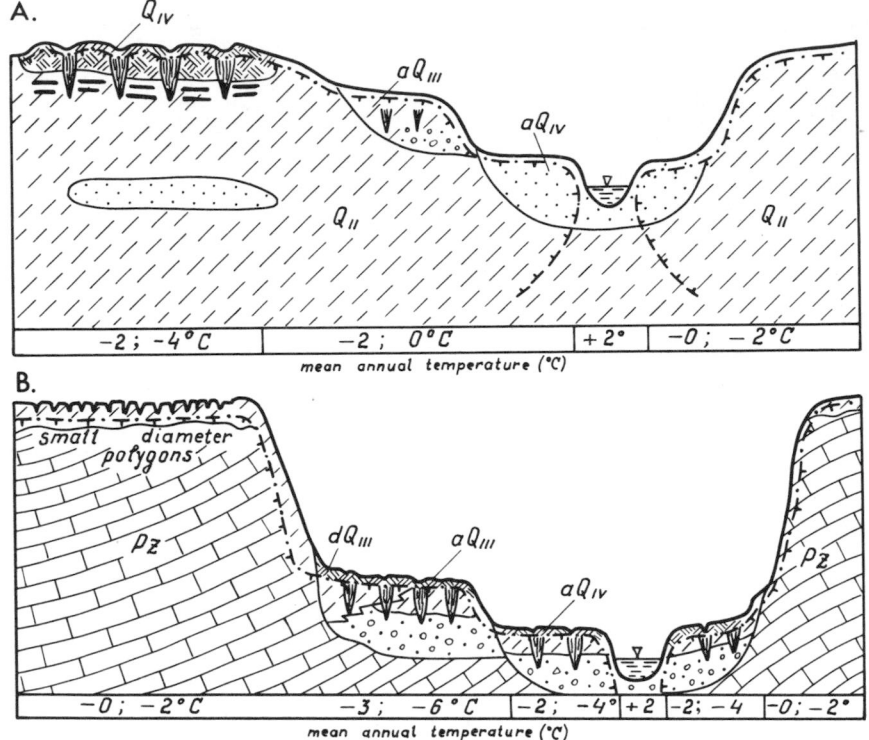

Fig. 9-6. The scheme of ice-wedge distribution in (A) West Siberia and (B) Middle and East Siberia.

occurrence, to raised peat masses mainly on the surface of high river terraces and on interfluves (figure 9-6). These peat bogs formed during the Holocene climatic optimum and were heaved owing to their freezing in the course of subsequent cooling. Under present-day climatic conditions in these regions, snow is blown from the surface of peat bogs in winter and they cool greatly. As a result, annual mean temperatures of the ground and minimum winter temperatures at the surface of peat bogs are lower than for other types of landscape. The shallow depth of seasonal thawing and survival of cracks in the thawed layer throughout summer aids in the persistent, active manifestation of the process. No clear-cut transition from ice to soil wedges is observed farther to the south.

The distribution of active wedges is at variance in Middle and East

Siberia. These territories are characterized by a severe, windless winter with a thin, uniform snow cover. The lowest ground temperatures are confined to valley bottoms. Here often lie diluvial and alluvial loams with a high ice content. Terrace surfaces are boggy. The ground thaws to a sufficiently large depth (1-1.5 m) so that the transition from ice to soil wedges is rather distinct.

Thus, in the north of both the European USSR and West Siberia, the southern boundary of the area of active ice wedges is drawn for peat bogs on flat-lying interfluves. The same boundary for Middle and East Siberia corresponds to ice wedges in alluvial and diluvial loams underlying valley bottoms. On high terraces, because of a severe continental climate, there are widely scattered soil wedges in networks of polygons. North of the southern boundary of the area of recently active ice wedges in sands and gravels, active frost cracking and the growth of ice wedges take place in practically all types of materials on all topographic features.

The southern boundary of the area of primordial, recently active soil wedges has not been established and is not shown on the map. The reason for this is that in continental areas, frost polygons with soil wedges in loam and sandy loam are of small diameter (1.5-3 m). They are comparable in size with desiccation polygons. The two processes, namely cracking due to frost contraction and desiccation, are frequently combined. The former acts in winter and the latter in summer. In both cases, small, morphologically similar soil veins are formed.

WEDGE NETWORKS

Frost polygons with active, primordial soil wedges and ice wedges overlain by a seasonally thawed layer more than 0.8-1.0 m thick have a raised centre and furrows above wedges. Frost polygons with active ice wedges, with the depth of seasonal thawing shallower than the above-indicated, are variously shaped. They may be flat, with narrow furrows above wedges; with raised centres, broad and deep lows above wedges; low centres and soil ridges above wedges (cf. Romanovskij, 1977). The map shows the boundary, south of which frost polygons with soil ridges (low-centre polygons) are practically nonexistent.

On denudation plateaux, tablelands, and mountains, where bedrock occurs near the surface and the seasonally thawed layer contains many fragments, sorted polygons are widespread. Both soil and ice wedges are associated with these polygons. Near the southern boundary of permafrost, many systems of sorted polygons with soil wedges appear to be of relict character. In this case, recently active frost cracking is nonexistent, but the wedges as such do develop. Owing to suffosion, silt and sand are forced upward onto the surface, frost heaving of rock fragments continues, and these fragments concentrate on the periphery of polygons and above wedges. On gentle mountain slopes and on denudation plateaux, such sorted polygons frequently serve as sources of block streams. Detailed investigations of block fields and block slopes carried out by A.I. Tyurin in Southern Yakutia (Romanovskij and Tyurin, 1979) have shown that the cover of coarse clastic material frequently conceals a polygonal net of wedge structures. Soil wedges consist here of clastic material without a dispersed groundmass. Such fields pass on slopes into systems of rather narrow block streams directed along soil wedges that are elongated down the slope (figure 9-7).

Geocryologists, palaeogeographers, and geologists are concerned with the origin of the "ice complex." This is composed of sandy loam and loam with high ice content and with thick syngenetic ice wedges, and is common on the lowlands of northern Yakutia and on the Central Yakutia Lowland, where it occupies broad terraces and vast interfluves. In other areas, ice complex is restricted to narrow strips of terraces in the valleys of rivers draining plains and mountains. When it was formed (in the upper Pleistocene), the climate was extremely cold, possibly the coldest for the entire Quaternary. At that time, frost cracking and the growth of ice and soil

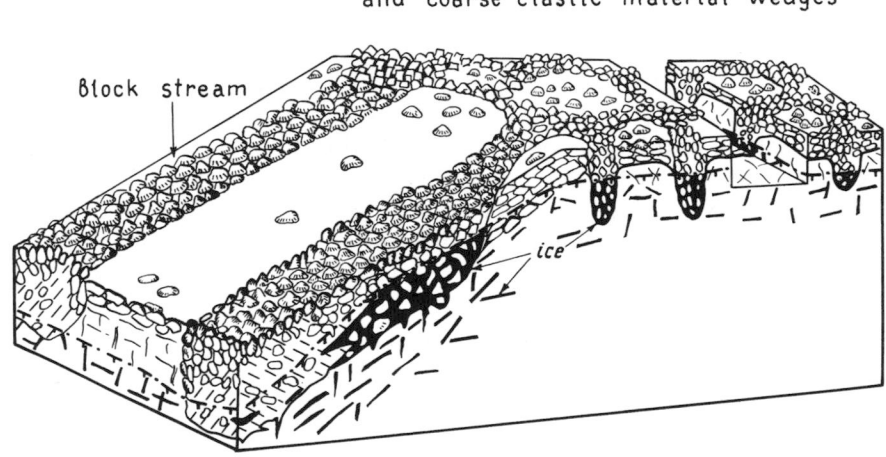

Fig. 9-7. Block field and block streams on denudation plateaux in Southern Yakutia.

wedges occurred in vast territories, at least throughout Siberia. The map shows areas of widespread ice complex and symbols denote areas of its local occurrence, mainly in river valleys, where it is restricted to the second and third terrace.

In conclusion, we direct attention to the fact that in recent years these deposits have been found in the northern Transbaikal south of the Stanovoy Upland, that is, much further south than had been assumed earlier. This suggests, first, that quite recently (20 to 10 thousand years ago) very severe permafrost conditions prevailed here and ground temperatures were below -6° or -8°C (now they are about -2°C or -3°C). Second, during the climatic optimum (8 to 4.5 thousand years ago), the ground did not thaw everywhere: patches of permafrost survived. We believe that the above-mentioned polygonal nets of soil wedges in the cryogenic, coarse clastic residual soils on the plateaux and mountains of Southern Yakutia and of the Northern Transbaikal were initiated at that cold time and still continue to develop. They affect block fields and streams and other exogenous phenomena. These observations allow us to elaborate our conceptions of the palaeogeography of the upper Pleistocene and Holocene ages in East Siberia.

REFERENCES

Romanovskij, N.N. 1977. Formirovaniye poligonal'no-zhil'nykh struktur [Formation of polygon-wedge structures]. Akad. Nauk SSSR. Sibirskoye Otdeliniye, Novosibirsk Izdatel'stvo 'Nauka': 215 pp.

Romanovskij, N.N. and Tyurin, A.I. 1979. Facial features of block streams of Southern Yakutia and the Northern Transbaikal. Vestnik Moskovskogo Universiteta, ser. 4, Geologiya, 1979 (4): 59-73.

10
PERIGLACIAL PROBLEMS

A.L. Washburn

Quaternary Research Center, The University of Washington
Seattle, WA 98195 U.S.A.

ABSTRACT

Some unresolved problems concerning periglacial processes and features and their use in environmental reconstructions are reviewed as a guide to future research.

RÉSUMÉ:
Problèmes périglaciaires

Des problèmes en suspens concernant des processus et des phénomènes périglaciaires ainsi que leur application dans l'interprétation des environnements anciens sont précisés en vue de guider les recherches à venir.

ПЕРИГЛЯЦИЯЛЬНЫЕ ПРОБЛЕМЫ

А. Л. ВАШБУРН

РЕЗЮМЕ

Рассматриваются некоторые нерешенные проблемы в области перигляциальных процессов и их особенностей. Применение таковых в реконструкции окружающей среды анализируется с целью указать дальнейший путь для исследовательской работы в будущем.

It is a great honour and pleasure to join in honouring Ross Mackay. In the following pages I shall stress what we do not know rather than what we do know, or think we know, about some periglacial processes and features. My intent is to erect some guideposts to research. I shall focus mainly and at some length on a number of fundamental problems concerning the exact nature of some of the processes and features, including questions regarding their distribution and environmental significance. Finally, I shall draw attention to related uncertainties affecting the use of relict or fossil periglacial features in environmental reconstructions. There are so many problems that my discussion is necessarily highly selective. It is also subjective in reflecting my greater familiarity with some questions than with others.

FROST HEAVING

Ice Segregation

Ice segregation and frost heaving are fundamental to a variety of periglacial features. Various concepts have been suggested to explain the suction potential (Anderson, 1971: 2-4) that draws water to a freezing zone in soil. An electric double layer (Chalmers and Jackson, 1970) and osmotic effects (Cass and Miller, 1959) have been cited as important factors. A capillary model (Everett, 1961) has been widely used but, as recently reviewed by Takagi (1980), this approach has some problems. Various thermodynamic models based largely on the Clapeyron or Clausius-Clapeyron relationships have been offered, which involve coupled heat and moisture transfer. These include an "adsorption-force" model (Takagi, 1980), a "segregation-potential" model (Konrad and Morgenstern, 1980; 1981), and a "hydrodynamic" model, presented by Michael Smith as part of this volume. Lorne Gold's contribution to the volume, in which he discusses "the ice factor in permafrost", bears importantly on thermodynamic approaches. O'Neill and Miller (1980: 657-659) have noted that capillary and thermodynamic models can be linked mathematically. That more information is needed regarding the coupled thermal, hydraulic, and solute fluxes accompanying ground freezing (Fukuda, 1980; National Academy of Sciences, 1974: 15-16), including water movement in frozen soils, is obvious. Contrary to earlier views, frozen soils can show significant permeability (Parmuzina, 1978, 1979, 1980; Perfect and Williams, 1980; P.J. Williams and Perfect, 1979; Zhestkova, 1978, 1980), and observations by Mackay and Lewis (1981) and more recent data (Mackay, 1981: 1675-1676) show that movement of porewater through permafrost can contribute significantly to frost heaving.

The exact nature of the freezing zone is a lively subject, as illustrated by the concept of secondary heaving involving the formation of ice behind a freezing front and the changes this ice can undergo (Miller, Loch, and Bresler, 1975: 1029-1030). That the character and distribution of ice in frozen soil — its cryogenic structure — can change under certain temperature gradients during cooling *or* warming has been demonstrated recently by Soviet scientists (Parmuzina, 1978, 1979, 1980; Zhestkova, 1978, 1980). However, the mechanics of this metamorphism remain problematic.

Other Frost Heaving Problems

There are many additional uncertainties relating to frost heaving that require further investigation. I will discuss some of these later in connection with patterned ground. Others include the precise mechanisms by which joint blocks are ejected from bedrock (Dyke, 1979, 1981) (figure 10-1) and stones from the soil (figure 10-2). Frost heaving of stones still leaves doubt about the relative importance of the frostheave and frostpush hypotheses [cf. Washburn, 1979 (reference subsequently cited as G): 86-91]

Fig. 10-1. Frost-heaved bedrock, Mesters Vig, Northeast Greenland. (cf. Washburn, 1969, figure 27: 51.)

— whether objects primarily are pulled from the ground by frost heaving of the surrounding soil or are pushed upward by ice accretion at their base — although I suspect the former prevails. The apparently sudden upfreezing of stones capped by silt and still-living vegetation is puzzling (figure 10-3) (cf. Washburn, 1969: 52,55, figure 30). Recent work by Ershov et al. (1978: 179, 1980: 171-174) and by Mackay (1981: 1674-1676) shows, surprisingly, that frost heaving is possible during thawing because of the redistribution of melt-water to build up ice lenses below the frost table.

FROST WEDGING

Frost wedging has long been regarded as the main physical weathering process in cold regions but, more recently, hydration shattering by sorption-desorption of water molecules has been advanced as being primarily responsible for the production of angular rock fragments (Dunn and Hudec, 1965: 115-138, 1966; Hudec, 1974; Konishchev, 1973, 1978a; Thorn, 1979; S.E. White, 1976a).

Counter arguments exist (Church, Stock, and Ryder, 1979: 391-393;

Fig. 10-2. Frost-heaved block in diamicton, Mesters Vig, Northeast Greenland. Scale given by trench shovel. (cf. Washburn, 1969, figure 28: 53.)

Fig. 10-3. Frost-heaved cobble with silt and vegetation capping, Mesters Vig, Northeast Greenland. Scale given by 16-cm rule. (cf. Washburn, 1969, figures 30-31: 55.)

Fahey and Gowan, 1979), and in general the predominance of angular fragments in cold regions as compared with temperate regions would appear to support the primary importance of frost wedging (G: 74). However, laboratory and theoretical considerations indicate the need for further research. Assuming that frost wedging is the primary process, the relative importance of the intensity of freezing versus the frequency of freeze-thaw cycles as factors requires additional work (cf. G: 75-76). So does frost wedging resulting from cycles confined to negative temperatures, since freezing of water in fine-grained soils occurs through a range of these temperatures (cf. Barsch, 1977b: 155; Fukuda, 1972; Konishchev, 1978b, 1980). Also, there is evidence that the weathering sequence of rock-forming minerals subjected to freeze-thaw cycles differs somewhat from other weathering sequences inasmuch as it seems capable of reducing quartz particles to smaller sizes than feldspar particles, and as small as 0.05-0.01 mm (Konishchev, 1978b, 1980; Mackay, Konishchev, and Popov, 1979). Frost wedging of bedrock in streams that freeze up has been suggested as being very important in stream erosion — Büdel's (1977: 60-61) ice-rind effect — but this has been challenged (cf. Bibus, 1975: 115; Bibus, Nagel, and Semmel, 1976: 39-41; Schunke, 1975: 188, 229; Semmel, 1976: 398; Stablein, 1977: 30-31).

There are still many questions related to freeze-thaw weathering of rocks that need to be resolved (cf. Aguirre-Puente, 1978, 1980).

FROST CRACKING

Frost cracking — the splitting of frozen ground as the result of thermal contraction at subfreezing temperatures — has been studied for many years. Nevertheless, details of the process continue to offer problems. One is presented by the rheology of frozen ground and its coefficient of thermal contraction, the latter varying with ice content and nature of the frozen soil. Elastic behaviour is frequently assumed, but deformation follows a power law involving time dependent creep (Andersland, Sayles, and Ladanyi, 1978: 216-263; Mackay, 1975: 1672-1674). This complicates the calculation of temperature requirements for frost cracking, thus handicapping the use of ice-wedge casts as palaeothermometers, although certain maximum mean annual air and ground temperatures can be inferred reasonably. The geotechnical importance of a field-tested flow law for ice-rich frozen ground has been reviewed for us in this volume by Norbert Morgenstern.

Determination of frost cracking as opposed to other causes of a crack pattern can be troublesome where additional cracking processes may be involved, especially desiccation, as noted by Nikolai Romanovskij in his paper (cf. G: 160-166). The problem is compounded when dealing with fossil features, such as ice wedge casts where apparently no single criterion of frost crack origin is diagnostic (Black, 1976: 11-12). The growth of ice wedges in frost cracks is well established but there are some uncertainties about the exact growth processes (Popov, 1978: 165-167; Popov and Katasonov, 1978: 720-721; Solomatin, 1979).

FROST SORTING

A variety of mechanisms exist whereby sorting effects are produced by frost action. I have already cited problems related to the upfreezing of stones, which is one such effect. Particle sorting during freezing has been little researched since Corte's (1961), 1962a, 1962b, 1962c, 1966a, 1966b) contributions and deserves renewed study.

Still other particle-by-particle sorting processes, such as differential thawing and eluviation, and perhaps even capillary action, need to be explored (cf. G: 118-119). In addition, there are many questions concerned with mass displacement whereby one kind of material is displaced *en masse* by another. In places it is far from clear whether artesian pressure, cryostatic pressure, changes in density and intergranular pressure, thawing pressure, differential volume changes during freezing and thawing, or perhaps still other processes are the most widespread and significant, or even the critical ones for a given feature (cf. G: 97-102).

A general question is the environmental significance of sorted forms. Some occur in warm arid regions (cf. Hunt and Washburn, 1966: B118) but are not likely to be confused with periglacial forms, which are the most common. Sorted nets less than 1 m in diameter (figure 10-4), or stripes less than 1 m in spacing, are widespread in cold nonpermafrost regions but sorted forms exceeding 2 m in dimension (figure 10-5) tend to be associated primarily with permafrost. More information is urgently needed for a quantitative assessment.

PATTERNED GROUND

General

Essentially, all the many varieties of patterned ground comprising nonsorted and sorted circles, polygons, nets, and stripes can be explained by the interaction of patterning, sorting, and slope processes, but in many situations the exact processes and their influence are far from clear. Yet the origin and distribution of patterned ground bear critically on both contemporary and palaeoenvironmental questions because of what the various forms can tell us. I have selected only certain varieties of patterned ground for discussion, along with involutions.

Circles

It is known that many circles, both nonsorted (figure 10-6) and sorted, involve a diapir-like movement from depth (figure 10-7). Hypotheses as to the cause(s) of the diapirism focus on various types of mass displacement (cf. G: 96-102, 167: Egginton, 1981: 300; Shilts, 1978; Veillette, 1980). More than one type may apply. Where diapirism is absent, simple local differential heaving may suffice (cf. G: 91, 167-168). To what extent polygonal cracking may be involved in initiating some forms that evolve into circles also requires research (cf. G: 161).

By the same token, the environmental implication of such circles is not clearly established. Most accounts of contemporary circles relate them to a permafrost environment but some apparently active circles occur where permafrost is lacking, as in South Georgia (Clayton, 1977: 94) and places in Iceland (Steche, 1933: 209).

Fig. 10-4. Small sorted polygons, High Valley, Denali Highway, Alaska. Scale given by 16-cm rule. (cf. Washburn, 1980, figure 39: 364.)

Fig. 10-5. Large sorted nets, edge of Greenland Ice Sheet, vicinity Thule, Greenland. (cf. Washburn, 1980, figure 21: 347.)

Fig. 10-6. Nonsorted circle, Mesters Vig, Northeast Greenland. Circle is 95-125 cm in diameter. (cf. Washburn, 1969, figures 41-42: 68.)

Fig. 10-7. Cross-section of nonsorted circle, Mesters Vig, Northeast Greenland. Note clayey sediment penetrating upward into sand. (cf. Washburn, 1969, figure 74 — Section D-D: 114-115.)

Fig. 10-8. Hummocks, Mesters Vig, Northeast Greenland. (cf. Raup, 1965; Washburn, 1979, figure 5.29: 146.)

Hummocks

Hummocks (figure 10-8) — a kind of vegetated nonsorted net — occur in a variety of forms, and the origin of some is problematic (cf. G: 146-147). Bordering cracks may be present (Drury, 1962: 36-37), often of uncertain origin, and Mackay (1979a) has suggested that certain hummocks are caused by thawing pressure which perhaps provides the net pattern and sets the stage (J.R. Mackay, personal communication, 1978). Other forms fail to show cracks despite careful search (Raup, 1965: 76). The demonstrable upward projection of fines in many forms argues for mass displacement but exactly what kind is often uncertain. Upfreezing of a stone can be a cause of hummocks (Dionne, 1966) but this clearly is not the case where a core stone is lacking.

Hummocks are locally prominent in polar regions (cf. G: 147), but they are primarily characteristic of subpolar and alpine environments and are not an indicator of permafrost. As yet no generally applicable maximum

mean annual air or ground temperature can be assigned to their development, although in Japan a mean annual air temperature of about 6°C has
been suggested (Koaze, Nogami, and Iwata, 1974: 177-180). A worldwide
survey relating the occurrence of hummocks to temperature and precipitation is essential if their environmental significance is to be examined fully.

Ice Wedge Polygons

The frostcrack origin of most ice wedge polygons (figure 10-9) is firmly
established, thanks to many researchers worldwide, especially Katasonov,
Popov, Romanovskiy, and others in the Soviet Union, and Black, Lachenbruch, Leffingwell, Mackay, and Péwé in North America (cf. G: 102-111,
133-141). Nevertheless, a number of uncertainties remains.

Questions include the interaction of factors controlling polygon size and
pattern, the former aspect being stressed recently by Tumel' (1979). The
exact manner in which marginal ridges develop during ice wedge growth in

Fig. 10-9. Ice wedge polygons, Tuktoyaktuk Peninsula, Northwest Territories, Canada. Photo
by Alfred Jahn.

low centre polygons is an example of things seeming to be somewhat different than commonly assumed (Mackay, 1980).

The presence of ice wedge polygons is of major import in land-use planning and engineering projects at high latitudes, and in environmental reconstructions based on ice wedge casts. Although ice wedge polygons are demonstrably permafrost forms, they require lower mean annual temperatures than those for permafrost alone; just how much lower depends upon the rheology of the frozen ground and upon the environmental conditions. Yet not only is the rheology itself subject to further investigation, as I noted earlier, but estimates of the temperature required for cracking depend upon the rate of temperature drop as well as upon the actual ground temperature (Lachenbruch, 1966: 65-66) as influenced by soil conditions and the insulating effect of snow and vegetation.

Consequently, the air and ground temperatures needed for permafrost cracking vary widely, with estimates of mean annual air temperatures ranging from -2° to -10°C. A widely quoted norm is Péwé's (1966a; 1966b) -6° to -8°C as a maximum mean annual air temperature for Alaskan conditions. Because of the variability of conditions worldwide, -5°C has been advanced as a more conservative figure, with air temperatures commonly being lower but rarely higher for well developed forms (G: 136-140, 313-319). This is still a crude rule of thumb and many more high latitude studies, such as those of Mackay (1963: 70, figure 28; 390, figure 2; cf. G: 139-140) and Péwé (1966a; 1966b), are needed for reliable regional comparisons.

Soil Wedge Polygons

Periglacial soil wedge polygons raise numerous questions, including terminological difficulties. Some Soviet investigators, for instance, restrict the term 'soil wedge' (or vein) and its equivalent, 'ground wedge' (or vein) — usually prefixing to the terms designations such as "original" or "primary" — to seasonally frozen ground in a nonpermafrost environment, or to the active layer above permafrost. Other investigators make no such distinction and regard sand wedges in permafrost as a type of soil wedge. The solution adopted here is to use the term soil wedge in a general sense to include all these possibilities and then to identify specific situations.

Although active soil wedges are widespread in permafrost environments, especially sand wedge types in sandy, vegetation-free areas of the Arctic and Antarctic, the extent to which active soil wedge polygons also occur in nonpermafrost environments is little known. The most that can be said is that the occurrence of most large, well developed soil wedge polygons is probably similar to that of ice wedge polygons, except for sand wedge types, which require a source of sand and, depending upon conditions, may imply aridity.

Periglacial Involutions

Periglacial involutions (figure 10-10) — the "... aimless deformation, distribution, and interpenetration of beds produced by frost action" (Sharp, 1942: 115) — pose problems. One is the difficulty of identifying periglacial involutions as such because of the variety of similar features that have no relation to frost action.

Some periglacial involutions are related to patterned ground, others not. A common interpretation of forms having a constant depth to their base is that they are permafrost features whose depth reflects the thickness of the active layer (cf. Poser, 1948: 54, 58). This may be true in places, but in others perhaps is due to stratigraphic conditions or to constant depth of freeze-thaw in the absence of permafrost. So little is known about periglacial involutions that a critical exploration of their origin, distribution, and significance is long overdue (cf. Corte, 1969: 141; G: 170-173).

STRING BOGS

String bogs (figure 10-11) — areas of peatland characterized by ridges of peat and vegetation interspersed with depressions that often contain shallow ponds — occur in considerable variety. Some string bogs have linear palsas as an important element of the string alignment (Åhman, 1976: 28, 1977: 41, 138; King, 1979). Although extensively studied, the origin of string bogs is not yet fully understood (cf. G: 174-176).

Questions also arise concerning the distribution of string bogs. In general they are most common in areas straddling the southern limit of discontinuous permafrost and thus south of most palsas, but some string bogs are also reported well within the continuous permafrost zone (G: 174-176). Again, superficially somewhat similar looking forms can occur that have no relation whatsoever to frost action (Greenwood, 1957; Macfadyen, 1950).

PALSAS

Palsas (figure 10-12) are mounds, circular or irregular, that contain perennial ice lenses that generally have peat as an important constituent and occur in bogs. Some forms regarded as palsas have a core of silt rather than peat, and rare forms may consist of silt alone (Åhman, 1976: 29-30). Maximum height is about 10 m. Several palsa types are recognized, includ-

Fig. 10-10. Periglacial involution, beach between Peris and Clydan, Wales, Great Britain. (cf. Watson, 1976: 92-96.)

Fig. 10-11. String bog, southeastern Labrador, Newfoundland, Canada. Scale given by trees. Photo by H.E. Wright, Jr.

Fig. 10-12. Palsa, vicinity Great Whale River, Quebec, Canada. Photo by R.J.E. Brown.

Fig. 10-13. Pingo, Wollaston Peninsula, Victoria Island, Northwest Territories, Canada. (cf. Washburn, 1956, figures 1-2, Pl. 13: 46.)

ing those whose individual relief is dynamic as opposed to those whose relief is the result of the thaw collapse of a former palsa plateau (cf. G: 176-179).

A number of problems can be cited. A troublesome one is terminology and whether there is a basic difference between palsa types (King, 1979: 43) or between large palsas of the dynamic type and some small pingos as discussed later (Åhman, 1977: 40, 138; Flemal, 1976: Lundqvist, 1969: 213).

The majority of typical palsas appear to have rather narrow environmental limitations. Palsas occur at mean annual air temperatures near 0°C in Iceland (Thorarinsson, 1951: 154-155), and below -1°C and 400 mm precipitation in Finland (Salmi, 1968: 182). Similarly, they are found below 0°C to -1°C and below 400 mm precipitation in Norway (Åhman, 1977: 143). In Sweden they occur below 0°C with a southern limit near -2° to -3°C (Lundqvist, 1962: 93; Rapp and Annersten, 1969: 67-69). In Canada they appear to be largely limited to the discontinuous permafrost zone (R.J.E. Brown, 1974; Railton and Sparling, 1973). That palsas or palsa-like forms can occur in the continuous zone is indicated by observations in Spitsbergen (Åkerman, 1980: 79; Salvigsen, 1977). Generally, however, depending on snow cover and other factors, the upper annual air temperature limit for the majority of typical palsas would appear to be about -1°C. A wider survey, including minimum temperature range, could be illuminating.

PINGOS

To discuss pingos (figure 10-13) in a presentation to Ross Mackay is not only bringing coals to Newcastle but bringing an inferior brand of coal in a leaky ship. I embark with trepidation.

Pingos are generally defined as large, perennial mounds containing cores of massive ice, as illustrated in a classic photo (figure 10-14) by Mackay. Some forms accepted as pingos are smaller than large palsas and some pingos are characterized by ice lenses rather than by a massive ice core (Mackay and Stager, 1966: 363-367). Peat, an important constituent of most palsas, is also present in some pingos (Porsild, 1938: 51-53). Perhaps the comparative lack of peat in pingos and its effect on the thermal regime account for pingos normally occurring in colder regions than palsas. If so, palsa growth may not require pre-existing permafrost but only concurrent aggradation of permafrost within the palsa itself (Matti Seppälä, personal communication, 1981.)

Thanks to Mackay, the origin of closed system pingos — or the hydrostatic system type (Mackay, 1979b: 8-9) — is well known. The origin of open-system pingos (Müller, 1959, 1963), or the hydraulic system type

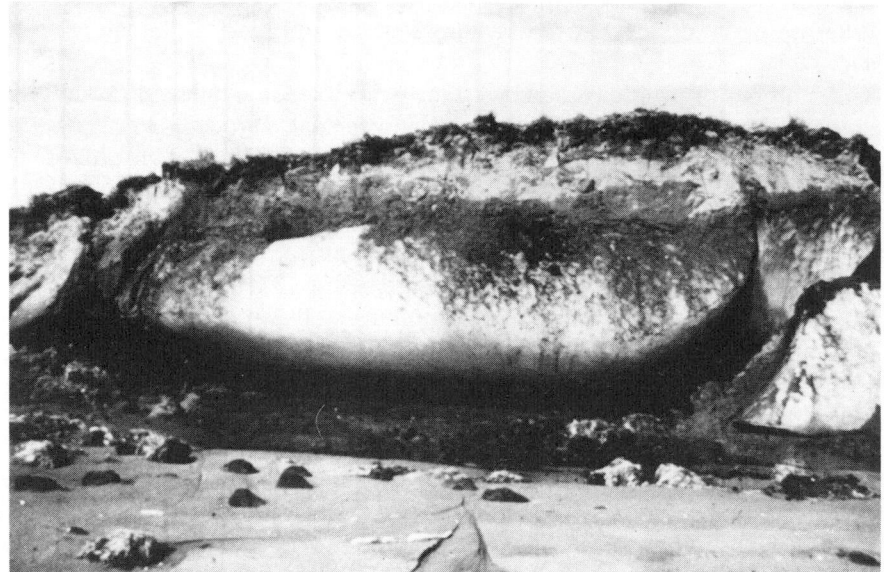

Fig. 10-14. Wave-cut cross-section of pingo showing ice core, McKinley Bay, c. 100 km northeast of Tuktoyaktuk, Mackenzie Delta area, Northwest Territories, Canada. Pingo was c. 90 m long and 7-10 m high. Photo by J.R. Mackay. (cf. Mackay, 1973a, figure 1: 980.)

Fig. 10-15. Gelifluction lobe, Hesteskoen, Mesters Vig, Northeast Greenland. (cf. Washburn, 1967, figure 40: 90.)

according to Mackay's recent usage (following Porsild, 1938: 47), is less well known (Holmes, Hopkins, and Foster, 1968: H32; J.R. Williams and van Everdingen, 1973: 440). It has been suggested that these pingo types converge (Pissart, 1970: 31-32).

Together with ice wedge casts, fossil pingos constitute one of the best evidences of former permafrost despite two main problems: identification of fossil pingos as such (Flemal, 1976) and their exact environmental significance. For hydrostatic pingos of northern Canada, Mackay (1978: 145) suggested a mean annual *ground* temperature of -5°C as an upper temperature limit for their occurrence. A maximum mean annual *air* temperature of -6°C has also been cited (Washburn, 1980: 360-361). For hydraulic pingos, Alaskan estimates include maximum mean annual air temperatures of -1°C (Péwé, 1969: 3) and -5.6°C (Holmes, Hopkins, and Foster, 1968: H7); Soviet estimates cite -3° to -4°C (Baulin, Dubikov, and Uvarkin, 1973: 15; 1978: 66).

MASS-WASTING FEATURES

Frost Creep and Gelifluction Deposits

Frost creep can be defined as "... the ratchetlike downslope movement of particles as the result of frost heaving of the ground and subsequent settling upon thawing, the heaving being predominantly normal to the slope and the settling more nearly vertical" (Washburn, 1967: 10). Gelifluction (figure 10-15) is solifluction associated with frozen ground (Baulig, 1956: 50-51), whereas solifluction, the more general term, is simply "... the slow flowing from higher to lower ground of masses of waste saturated with water ..." (Andersson, 1906: 95-96)

Frost creep, and probably soil creep generally, is usually accompanied by a retrograde component whose cause and quantitative effect are poorly understood. Although the cause is commonly ascribed to cohesion, details are lacking despite the fact that the effect was reported long ago (Davison, 1889). With respect to gelifluction, questions include the detailed mechanism of movement, whether primarily by shearing as in some clayey soils (cf. Hutchinson, 1974: 439), by blocklike movement (Mackay, 1981; Rein and Burrous, 1980), by small scale gliding along thawing ice lenses (Beskow, 1930: 624-625; Carson and Kirkby, 1972: 277), or simply by saturated flow, probably aided by gliding along ice lenses (cf. G: 205). Quite possibly the mechanisms and their importance vary with environmental conditions.

Although mechanically separate processes, frost creep and gelifluction combine their effect and so it is difficult to determine which is the more

Fig. 10-16. Cryoplanation terrace, Indian Mountain, Alaska. Photo by T.L. Péwé. (cf. Reger and Péwé, 1976: 107-108.)

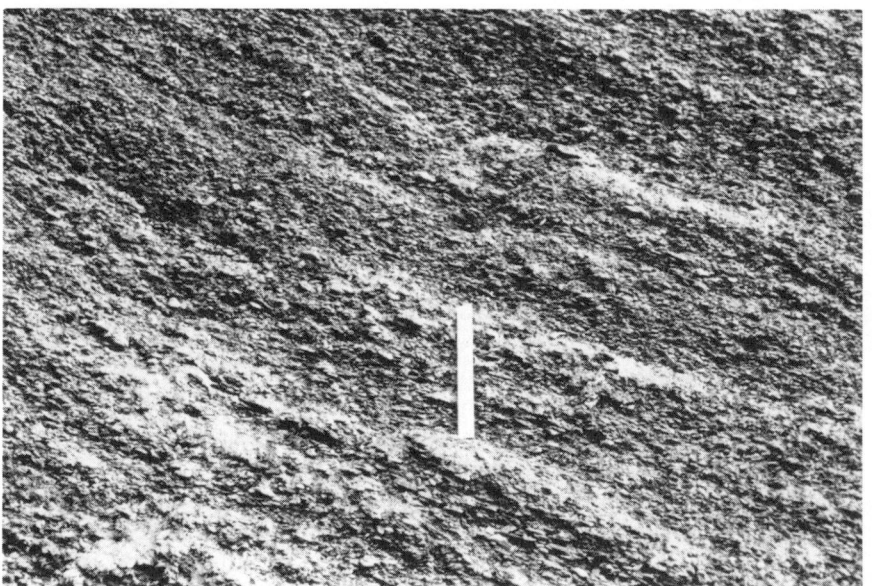

Fig. 10-17. Grèzes litées near Cemmaes, Wales, Great Britain. Scale given by 16-cm rule. (cf. Washburn, 1979, figure 8.1: 245.)

important at a given site or over a region. Possibly thermally induced soil creep, as distinct from frost creep, also contributes to the total movement as suggested by Mackay (1980, 1981). In any event, with respect to frost creep and gelifluction, it is uncertain whether each process produces distinctive deposits.

Another problem, critical to environmental interpretations, is the degree to which frost creep and gelifluction deposits can be distinguished from soil creep and solifluction deposits in nonperiglacial climates. The angularity of material suggestive of frost wedging is a common criterion but would not be forceful if hydration shattering by sorption-desorption of water molecules is capable of producing angular fragments as mentioned earlier.

Block Fields, Block Slopes, and Block Streams

Block fields and block slopes are broad areas covered with angular rocks, block slopes being on gradients exceeding 10°; block streams are similar but relatively narrow, linear deposits oriented downslope. Normally, unlike scree, all these deposits lack a cliff at their head, and all have long attracted attention and stimulated much debate as to origin and environmental significance (cf. G: 219-223). The predominance of mechanical weathering, especially frost wedging, is commonly believed to be critical to their origin except by those who support sorption-desorption of water molecules as being more important. Questions include the extent to which the deposits have developed *in situ*, moved significantly, or have been modified by eluviation of fines; and, if they have moved, the mechanics of movement.

Rock Glaciers

A rock glacier is "… a tongue-like or lobate body usually of angular boulders that resembles a small glacier, generally occurs in high mountainous terrain, and usually has ridges, furrows, and sometimes lobes on its surface, and has a steep front at the angle of repose" (Potter, 1972: 3027; cf. Potter, 1969: 1). There is a variety of forms, and their origin can be manifold.

Some rock glaciers evolve from true glaciers as shown by a core of glacier ice. Others appear to be unrelated to glaciers. The distinction is important because active rock glaciers are now generally regarded as permafrost features (cf. G:230). Inactive rock glaciers of nonglacial origin therefore indicate former zonal permafrost, whereas glacier-derived forms would be azonal in the sense of being able to reach altitudes below the normal permafrost limit. Distinguishing between these types, including active forms, can be problematic in places.

The mechanics of rock glacier movement is not well established, in part because of inadequate information concerning internal characteristics, such as grain size and ice content. The flow properties of ice-rock mixtures led Whalley (1974: 6-8, 20-28; cf. Whalley, 1979) to postulate a core of glacier ice as necessary for movement. However, a high ice content can be nonglacial (cf. Barsch, 1977a: 235; Fisch, Fisch, and Haeberli, 1977: 245). Movement has also been attributed to the presence of interstitial ice alone (P.G. White, 1976; S.E. White, 1976b: 89) or of ice lenses, the latter facilitating movement by eliminating grain-to-grain contact between rock components (cf. G: 229). Movement by basal shearing is another possibility (Wahrhaftig and Cox, 1959: 427-428; cf. Haeberli, 1979).

NIVATION FEATURES

Hollows and Benches

The concept of nivation, "... the localized erosion of a hillside by frost action, mass-wasting, and sheetflow or rillwork of meltwater at the edges of, and beneath lingering snowdrifts" (G: 325), has been with us for almost a century (cf. Matthes, 1900). Nevertheless, quantitative studies are conspicuously few. Thorn's (1974; 1976) work in the Colorado Front Range suggests that some traditional concepts need to be checked; for instance, that freeze-thaw cycles are more frequent adjacent to snowbanks than elsewhere, that nivation initiates hollows as opposed to modifying them, and that it can lead to development of cirques in favourable locations.

Cryoplanation Terraces

Nivation and gelifluction are believed to be largely responsible for cryoplanation terraces (figures 10-16), which are "... hillside or summit benches that are cut in bedrock, predominantly transect lithology and structure, and are confined to cold climates" (G: 237). They are reported to be widespread (Demek, 1969) but most descriptions are of relict forms, and there is suspicion that many features classed as cryoplanation terraces are really lithologic or structural benches. This is true of some occurrences in Alaska (Miotke, 1979: 124). Indeed, it remains to be demonstrated that nivation and gelifluction can fully explain their origin independently of lithologic or structural influence.

According to Demek (1969: 57), permafrost facilitates their origin but is not a necessary condition, so that the mean annual air and ground temperatures for their development could exceed 0°C. On the other hand, according to Reger and Péwé (1976: 107-108) cryoplanation terraces imply a very rigorous climate, specifically a former mean annual air temperature of about -12°C for an Alaskan occurrence they studied. The entire subject of cryoplanation terraces demands critical examination.

GRÈZES LITÉES

Grèzes litées (figure 10-17) are "... bedded slope deposits of angular, usually pebble-size rock chips and interstitial finer material, in which the bedding is manifested by more or less regularly repeated alternation of grain-size characteristics" (G: 244). Although such deposits are frequently regarded as periglacial in the European literature, they have been little studied outside Europe, and their exact nature, origin, and environmental significance are uncertain.

PERMAFROST

Age

A pervasive question relating to the origin of contemporary and relict permafrost is that of age. That permafrost is the result of postglacial climate is easily demonstrable where it ocurs in postglacial deposits or in older deposits first exposed to freezing in postglacial time. The age of the deposits that have remained frozen since earlier times is much more difficult to determine.

Existing permafrost in Siberia is reported to be as old as Lower Pleistocene (Katasonov, 1977) but the specific evidence is not elaborated. In the Yana-Indigirka lowland, syngenetic ice wedges (cf. G: 108, 110) demonstrate the continuing presence of permafrost during the deposition of sediments dated as Middle Pleistocene (Konishchev and Kartashova, 1972). That some permafrost in northern Canada is older than 40,000 years has been demonstrated by Mackay (1976a), and in Alaska some ice wedges are clearly older than 14,000 years or possibly as old as about 30,000 years (Sellman, 1967; 1972). However, the age of much existing permafrost is open to question pending more information on permafrost thicknesses, associated thermal profiles, absolute dating, and palaeontologic and stratigraphic data.

Massive Ice

One of the intriguing problems concerning permafrost is the origin of massive ice beds (figure 10-18); that is, ice layers over 2 m thick and 10 m in shortest diameter and having an ice content of at least 250 per cent on an

Fig. 10-18. Massive ice underlying stony clay with ice wedge, Tuktoyaktuk Peninsula, Northwest Territories, Canada. Massive ice is c. 6.0 m thick; stony clay, c. 4.5 m thick. Photo by J.R. Mackay. (cf. Mackay, 1972, figure 9: 9.)

ice-to-dry-soil weight basis (Mackay, 1973b: 223). Massive ice layers up to 50 m or more thick and up to 2 km or more in horizontal extent are known in Canada (Mackay, 1973b), and massive ice up to 80 m thick has been reported in Siberia (Gerasimov and Markov, 1968: 14; Grave, 1968a: 53, 1968b: 9; cf. Shumskiy, 1964: 199).

A number of suggestions for the origins of massive ice have been offered. They include buried glacier ice, buried icings, coalescing ice wedges (Gerasimov and Markov, 1968: 14; Grave, 1968a: 53; 1968b: 9; Shumskiy, 1964; 199), injection ice (Shumskiy and Vtyurin, 1966: 108, 110), and/or segregation ice (Mackay, 1971; 1972: 15-19; 1973b; 1976b; Mackay and Black, 1973: 187, 189; cf. Pissart and French, 1976). Probably all these origins apply to some degree, and criteria exist on the nature of the ice and sediments to help distinguish between some of them (cf. G: 42-49). Presumably origins such as buried glacier ice and buried icings are unusual except locally, at least in North America. More information is needed on the coalescence of ice wedges as an explanation, since lateral growth of wedges may be self limiting (Black, 1976: 6-7; Gasanov, 1973; 1978). Mackay's concept of growth of segregation ice aided by porewater expulsion provides an attractive explanation, which further research may well prove to be the most general.

Thermokarst

The development of thermokarst — thaw features in permafrost — offers a number of problems. Thermokarst depressions due to thawing ground ice are similar to glacial kettle holes, but the interpretation is vastly different in that the former are evidence of thawing permafrost, whereas the latter indicate former glaciation but not necessarily perennially frozen soil. A palaeoenvironmental problem is to distinguish between the two types of depressions. If the assumption is made that thermokarst depressions vanish upon the complete thawing of permafrost, the problem also disappears, but in this event the assumption remains to be proved.

The oriented lakes of the Liverpool Bay area east of the Mackenzie Delta, which were studied by Mackay (1956; 1963: 46-55), some lakes in western Baffin Island as well as in the Old Crow Plain, Yukon Territory, and the oriented lakes of the Coastal Plain of Alaska are all examples of thaw lakes whose pattern has caused considerable debate. The effects of wind-generated lake currents eroding at right angles to the dominant wind are now generally accepted as the principal cause of the pattern, but in detail Mackay's (1963: 54-55) conclusion that "... the precise mechanism of lake orientation remains unexplained" still applies. At least locally, bedrock structure as a factor does not appear to be completely eliminated (cf. G: 272-273). It is noteworthy that basement lineaments underlying unconsolidated sediments as much as 3000 m thick may explain the pattern for a large area of oriented lakes in Bolivia (Plafker, 1964).

Ecology of Permafrost

The ecology of permafrost, especially its reaction to environmental change, natural or human, is becoming increasingly important (J. Brown and Grave, 1979). Wherever permafrost exists, development activities require consideration of their effects upon it. How to mitigate thermokarst development or other, often complex, impacts concern the architect, engineer, and land-use planner. The questions are many and difficult, and they increasingly demand an integrated, problem-oriented approach involving expertise from numerous disciplines.

Carbon Dioxide – Permafrost Interactions

Temperature increase from a doubling of atmospheric carbon dioxide consequent upon human activities has been projected to be three to four times greater at the poles than in the tropics (MacDonald *et al.*, 1979: 3), amounting to estimated increases of 2°C to more than 7°C at high latitudes by the year 2050 (Bach, 1979: 67, Table 6, 74). Such increases would lead to appreciable increases in permafrost temperature and, especially in marginal areas, to a large increase in the thickness of the active layer or to the complete disappearance of permafrost.

Some permafrost contains appreciable organic matter whose thawing and oxidation would release further carbon dioxide which would enhance the warming effect. On the other hand there is evidence that the recent trend in polar climates is toward oscillatory cooling (Kukla *et al.*, 1977), and if permafrost increases it would become a carbon sink, thereby favoring additional cooling rather than warming. Thus, the amount of carbon and the carbon flux in the tundra and the boreal forest affect estimates of CO_2 - induced climatic change (Revelle *et al.*, 1980: 17-111).

From several viewpoints, therefore, potential interactions between carbon dioxide and permafrost warrant priority monitoring and research.

ENVIRONMENTAL RECONSTRUCTIONS

Key Features

Many uncertainties arise in using periglacial features as palaeoclimatic indicators in environmental reconstructions (Washburn, 1980). Of prime importance is the correct identification of the features, often a task in itself. Moreover, the features must be dated to within a sufficiently narrow time interval to be useful for the purpose at hand.

The key periglacial features for palaeoenvironmental reconstruction are those indicating former permafrost. They comprise relict permafrost, ice wedge cast polygons, large well-developed fossil soil wedge polygons, and fossil occurrences of palsas, pingos, thermokarst features, and some (perhaps all) cryoplanation terraces. With respect to climatic inferences from the features, generally only the maximum mean annual air or ground temperature can be estimated reasonably without more data than are at present available.

Climatic Estimates

Especially with periglacial evidence, serious difficulties result from factors that influence the effect of climate on ground temperature, among the most important being the damping effect of snow and vegetation. Thus, ground temperatures are normally somewhat warmer than air temperatures, on average by 3.3°C in northern Canada but with considerable regional variance (Judge, 1973: 38). Yet because of the scarcity of data regarding ground temperatures, air temperatures are usually cited, particularly mean annual estimates. Seasonal temperatures and freezing and thawing indices can be more meaningful. An interesting scheme for plotting the distribution of periglacial features with respect to mean annual air temperature, freezing and thawing indices, and permafrost was presented by Harris (1981). Precipitation indications are also provided by some periglacial features but less precisely than temperature estimates.

The reconstruction of past climates through periglacial evidence requires extreme caution, but also holds promising rewards. To aid interpretations, high priority should be given to critical palaeoenvironmental studies comparing the periglacial evidence with that derived from other types of proxy data.

Conflicts

The literature contains a number of conflicting interpretations of palaeofeatures that occur in the same general area and are of the same (sometimes presumed) date. Some illustrations are:

(1) New England features accepted as ice wedge casts by Schafer and Hartshorn (1965: 124; cf. Hartshorn and Schafer, 1965: 17) but rejected by Black (1966: 23; 1980);

(2) New Jersey Coastal Plain features accepted as periglacial by Wolfe (1953; 1956) but discounted by Péwé (R.J.E. Brown and Péwé, 1973: 91; Péwé, 1973: 22; cf. Sugden, 1977: 28-29), although more recently brought again into question by the work of Watts (G: 304; cf. Walters, 1978);

(3) The Mima Mounds of western Washington (Péwé, 1948; G: 169-170, 304);

(4) The presence or absence of fossil permafrost features in New Zealand (Harris, 1976; Soons, 1962: 83, 87).

CONCLUSION

There are numerous periglacial processes and features whose exact nature remains to be established. Accordingly, their use as palaeoclimatic indicators is handicapped, even though certain approximations are possible. The research challenges before us are exciting — the more so because of the leadership, the innovative methodology, and the stimulating insights that have been provided by Ross Mackay.

ACKNOWLEDGEMENTS

The writer is very grateful to Bernard Hallet and David Hopkins for helpful suggestions.

REFERENCES

Aguirre-Puente, Jaime. 1979. État actuelle des recherches sur le gel des roches et des matériaux de construction. *In* 3rd Int. Conf. on Permafrost, Edmonton, Canada, 10-13 July, 1978. Proc. *1*. Ottawa NRC Canada Pub. *16529*: 599-607.

Aguirre-Puente, Jaime. 1980. Present state of research on the freezing of rocks and construction materials [État actuelle des recherches sur le gel des roches et des matériaux de construction]. *In* 3rd Int. Conf. on Permafrost, Edmonton, Canada, 10-13 July, 1978. English translations of invited theme papers, Part 2. Ottawa, NRC Canada Pub. *18119*: 117-132.

Åhman, Richard. 1976. The structure and morphology of minerogenic palsas in northern Norway. Biuletyn Peryglacjalny *26*: 25-31.

Åhman, Richard. 1977. Palsar i Nordnorge. En studie av palsars morfologi, utbredning och klimatiska förutsättningar i Finnmarks och Troms fylke. Lunds Universitets Geog. Inst., Medd. Avh. *78*: 165 pp.

Åkerman, Jonas. 1980. Studies on periglacial geomorphology in West Spitsbergen. Lunds Universitets Geog. Inst., Medd. Avh. *89*: 297 pp.

Andersland, O.B., Sayles, F.H., jr. and Ladanyi, B. 1978. Mechanical properties of frozen ground. *In* Andersland, O.B. and Anderson, D.M., eds. Geotechnical engineering for cold regions. N.Y., McGraw-Hill: 216-275.

Anderson, D.M. 1971. Remote analysis of planetary water. U.S. Army Corps of Engineers, Cold Regions Res. and Eng. Lab., Research Rep. *274*: 19 pp.

Andersson, J.G. 1906. Solifluction, a component of subaerial denudation. J. Geology *14*: 91-112.

Bach, Wilfrid, 1979. Short-term climatic alterations caused by human activities: status and outlook. Progress in Phys. Geography *3*: 55-83.

Barsch, Dietrich. 1977a. Nature and importance of mass-wasting by rock glaciers in alpine permafrost environments. Earth Surface Processes *2*: 231-245.

Barsch, Dietrich. 1977b. Eine Abschätzung von Schuttproduktion und Schutt-transport im Bereich activer Blockgletscher der Schweizer Alpen. Z. Geomorphologie N.F., Supp. *28*: 148-160.

Baulig, Henri. 1956. Pénéplaines et pédiplaines. Soc. belge d'études géog. *25*: 25-58.

Baulin, V.V., Dubikov, G.I. and Uvarkin, YuT. 1973. Osnovnyye cherty stroyeniya i razvitiya vechnomerzlykh tolshch Zapadno-Sibirskoy ravniny. *In* Akademiya Nauk SSSR, Sektsiya Nauk o Zemle, Sibirskoye Otdeleniye, II Mezhdunavodnaya Konferentsiya po Merzlotovedeniyu, Doklady i soob-shcheniya *2* (Regional'naya geokriologiya). Yakutsk, Yakutskoye Knizhnoye Izdatel'stvo: 10-17.

Baulin, V.V., Dubikov, G.I. and Uvarkin, YuT. 1978. The main features of the structure and development of the permafrost of the west Siberian Plain [Osnovnye cherty stroyeniya i razvitiya vechnomerzlykh tolshch Zapadno-Sibirskoy ravniny]. *In* Sanger, F.J., ed. Permafrost 2nd Int. Conf., Yakutsk, USSR, 13-28 July, 1973: USSR Contribution. Washington, D.C., NAS (U.S.A.): 63-67.

Beskow, G. 1930. Erdfliessen und Strukturböden der Hochgebirge im Licht der Frosthebung. Geol. Fören. Stockholm, Förh. *52*: 622-638.

Bibus, Erhard. 1975. Geomorphologische Untersuchungen zur Hang- und Talentwicklung im zentralen West-Spitzbergen. Polarforschung *45*: 102-119.

Bibus, Erhard, Nagel, G. and Semmel, A. 1976. Periglazialer Reliefformung im zentralen Spitzbergen. Catena *3*: 29-44.

Black, R.F. 1976. Periglacial features indicative of permafrost: ice and soil wedges. Quat. Research *6*: 3-26.

Black, R.F. 1980. Evidence of permafrost during the latest Wisconsinan ice advance and retreat in Connecticut. *In* Amer. Quat. Assoc., 6th Bienn. Mtg, Orono, Me., 18-20 Aug., 1980, Abstracts and Program. Orono, Univ. Maine, Inst. for Quat. Studies: 25.

Brown, Jerry and Grave, N.A. 1979. Physical and thermal disturbance and protection of permafrost. *In* 3rd Int. Conf. on Permafrost, Edmonton, Canada, 10-13 July, 1978. Proc. *2*. Ottawa, NRC Canada Pub. *16529*: 51-91.

Brown, R.J.E. 1974. Some aspects of airphoto interpretation of permafrost in Canada. NRC Canada, Div. Building Research Tech. Paper *409*: 20 pp.

Brown, R.J.E. and Péwé, T.L. 1973. Distribution of permafrost in North America and its relationship to the environment: a review, 1963-1973. *In* Permafrost 2nd Int. Conf., Yakutsk, USSR, 13-28 July, 1973: North American Contribution, Washington, D.C., NAS(U.S.A): 71-100.

Büdel, Julius. 1977. Klima-Geomorphologie. Berlin and Stuttgart, Gebrüder Bornträger: 304 pp. (English translation by L. Fischer and D. Busche, 1982. Climatic Geomorphology by Julius Büdel, Princeton, N.J., Princeton Univ. Press. 443 pp.)

Carson, M.A. and Kirkby, M.J. 1972. Hillslope form and process. Cambridge University Press. 475 pp.

Cass, L.A. and Miller, R.D. 1959. Role of the electric double layer in the mechanism of frost heaving. U.S. Army Snow, Ice and Permafrost Research Establ., Research Rep. *49*: 15 pp.

Chalmers, Bruce and Jackson, K.A. 1970. Experimental and theoretical studies of the mechanism of frost heaving. U.S. Army Corps of Engineers, Cold Regions Research and Eng. Lab., Research Rep. *199*: 22 pp.

Church, Michael, Stock, R.F. and Ryder, J.M. 1979. Contemporary sedimentary environments on Baffin Island, N.W.T., Canada: debris slope accumulations. Arctic and Alpine Research *11*: 371-402.

Clayton, R.A.S. 1977. The geology of northwestern South Georgia: I. Physiography. British Antarctic Surv., Bull. *46*: 85-98.

Corte, A.E. 1961. The frost behavior of soils: laboratory and field data for a new concept — I, vertical sorting. U.S. Army Corps of Engineers, Cold Regions Research and Eng. Lab., Research Rep. *85*(1): 22 pp.

Corte, A.E. 1962a. Vertical migration of particles in front of a moving freezing plane. J. Geophys. Research *67*: 1085-1090.

Corte, A.E. 1962b. Relationship between four ground patterns, structure of the active layer, and type and distribution of ice in the permafrost. U.S. Army Corps of Engineers, Cold Regions Research and Eng. Lab., Research Rep. *88*: 79 pp.

Corte, A.E. 1962c. The frost behavior of soils: laboratory and field data for a new concept — II, horizontal sorting. U.S. Army Corps of Engineers, Cold Regions Research and Eng. Lab., Research Rep. *85*(2): 20 pp.

Corte, A.E. 1966a. Experiments on sorting processes and the origin of patterned ground. *In* Permafrost Int. Conf., Lafayette, Indiana, 11-15 Nov., 1963, Proc. NAS/NRC(U.S.A.) Pub. *1287*: 130-135.

Corte, A.E. 1966b. Particle sorting by repeated freezing and thawing. Biul. Peryglacjalny *15*: 175-240.

Corte, A.E. 1969. Geocryology and engineering. *In* Varnes, D.J. and Kiersch, George, eds. Revs. in Eng. Geology II. Boulder, Col., Geol. Soc. Amer.: 119-185.

Davison, Charles. 1889. On the creeping of the soilcap through the action of frost. Geol. Mag. (Great Britain), decade 3, v. *6*: 255-261.

Demek, Jaromír. 1969. Cryoplanation terraces, their geographical distribution, genesis and development. Československé Akademie Véd Rozpravy, Řada Matematickych a Přírodních Véd: *79*(4): 80 pp.

Dionne, J.-C. 1966. Un type particulier de buttes gazonnées. Rev. Géomorphologie Dynamique *16*: 97-100.

Drury, W.H. jr. 1962. Patterned ground and vegetation on southern Bylot Island, Northwest Territories, Canada. Harvard Univ., Gray Herbarium Contrib. *190*: 111 pp.

Dunn, J.R. and Hudec, P.P. 1965. The influence of clays on water and ice in rock pores (2). N.Y. State Dept. Public Works, Phys. Research Rep. *RR 65-5*: 149 pp.

Dunn, J.R. and Hudec, P.P. 1966. Frost deterioration: ice or ordered water? (abst.). Geol. Soc. Amer., Spec. Paper *101*: 256.

Dyke, L.D. 1979. Bedrock heave in the central Canadian Arctic. Geol. Surv. Canada, Paper *79-1A*: 241-246.

Dyke, L.D. 1981. Bedrock heave in the central Canadian Arctic. Geol. Surv. Canada. Paper *81-1A*: 157-167.

Egginton, P.A. 1981. The impact of disturbance on mudboil activity, north Henik Lake, District of Keewatin. Geol. Surv. Canada, Paper *81-1A*: 299-303.

Ershov, E.D. *et al.* 1978. Migratsiya vlagi, strukturoobrazovanie i shlirovoe l'dovydelenie v promerzayushchikh i ottaivayushchikh glinistykh gruntakh. *In* 3rd Int. Conf. on Permafrost, Edmonton, Canada 10-13 July, 1978, Proc. *1*. Ottawa, NRC Canada Pub. *16529*: 175-180.

Ershov, E.D. *et al.* 1980. Water migration, formation of texture and ice segregation in freezing and thawing clayey soils. *In* 3rd Int. Conf. on Permafrost, Edmonton, Canada, 10-13 July, 1978. English translations of invited theme papers, Part I. Ottawa, NRC Canada, Pub. *18119*: 159-175.

Everett, D.H. 1961. The thermodynamics of frost damage to porous solids. Faraday Soc. Trans. *57*: 1541-1551.

Fahey, B.D. and Gowan, R.J. 1979. Application of the sonic test to experimental freeze-thaw studies in geomorphic research. Arctic and Alpine Research *11*: 253-260.

Fisch, W., Sr., Fisch, W., jr. and Haeberli, W. 1977. Electrical DC resistivity soundings with long profiles on rock glaciers and moraines in the Alps of Switzerland. Z. Gletscherkunde u. Glazialgeol. *13*: 239-260.

Flemal, R.C. 1976. Pingos and pingo scars: their characteristics, distribution, and utility in reconstructing former permafrost environments. Quat. Research *6*: 37-53.

Fukuda, Masami. 1972. Ganseki nai no mizu no toketsu ni tsuite — II (Freezing-thawing process of water in pore space of rocks — II). Low Temperature Science, Ser. A, *30*: 183-189.

Fukuda, Masami. 1980. Experimental studies of coupled heat and moisture transfer in soils during freezing. Cold Regions Science and Technology *3*: 223-232.

Gasanov, SH.SH. 1973. Mekhanizm samoregulirovaniya predel'nykh razmerov ledyanykh zhil. *In* Akademiya Nauk SSSR, Sektsiya Nauk o Zemle, Sibir-skoye Otdeleniye, II Mezhdunarodnaya Konferentsiya po Merzlotovedeniyu, Doklady i soobshcheniya *3* (Genezis, sostav i stroyeniye merzlikh toshch i podzemnyye l'dy). Yakutsk, Yakutskoye Knizhnoye Izdatel'stvo: 65-69.

Gasanov, SH.SH. 1978. Self-regulating mechanism of size-limited ice wedges [Mekhanizm samoregulirovaniya predel'nykh razmerov ledyanykh zhil]. *In* Sanger, F.J., ed. Permafrost 2nd Int. Conf., Yakutsk, USSR, 13-28 July, 1973: USSR Contribution. Washington, D.C., NAS (U.S.A.): 195-197.

Gerasimov, I.P. and Markov, K.K. 1968. Permafrost and ancient glaciation. Defence Research Bd. Canada, Transl. *T499R*: 11-19.

Grave, N.A. 1968a. Merzlyye tolshchi zemli. Priroda *1968(1)*: 46-53.

Grave, N.A. 1968b. The earth's permafrost beds [Merzlyye tolshchi zemli]. Defence Res. Bd. Canada, Trans. *T499R*: 1-10.

Greenwood, J.E.G.W. 1957. The development of vegetation patterns in Somaliland Protectorate. Geog. J. *123*: 465-473.

Haeberli, Wilfried. 1979. Holocene push-moraines in alpine permafrost. Geog. Annal. *61A*: 43-48.

Harris, S.A. 1976. Loess wedges on the Banks Peninsula, South Island, New Zealand. Can. Assoc. Geographers Annual Meeting, Univ. Laval, 23-27 May, 1976, Prog. and Résumés: 46-49.

Harris, S.A. 1981. Distribution of zonal permafrost landforms with freezing and thawing idices. Erdkunde *35*: 81-90.

Hartshorn, J.H. and Schafer, J.P., organizers. 1965. New England. *In* Int. Assoc. for Quaternary Res. (INQUA) 7th Cong., Boulder, Col., Guidebook for Field Conference A, New England - New York State: 5-38.

Holmes, G.E., Hopkins, D.M. and Foster, H.L. 1968. Pingos in central Alaska. U.S. Geol. Surv., Bull. *1241-H*: 40 pp.

Hudec, P.P. 1974. Weathering of rocks in Arctic and sub-Arctic environments. *In* Aitken, J.D. and Glass, D.J., eds. Canadian Arctic Geology. Geol. Assoc. Canada - Can. Soc. Petroleum Geologists, Symposium on the geology of the Canadian Arctic, Saskatoon, 24-26 May, 1973. Proc.: 313-335.

Hunt, C.B. and Washburn, A.L. 1966. Patterned ground. *In* Hunt, C.B. et al, Hydrologic basin, Death Valley, California. U.S. Geol. Surv., Prof. Paper *494-B*: B104-B133.

Hutchinson, J.N. 1974. Periglacial solifluxion: an approximate mechanism for clayey soils. Géotechnique *24*: 438-443.

Judge, A.S. 1973. Deep temperature observations in the Canadian North. *In* Permafrost 2nd Int. Conf. Yakutsk, USSR, 13-28 July, 1973: North American Contribution, Washington, D.C., NAS(U.S.A.): 35-40.

Katasonov, E.M. 1977. Cryolithogenic deposits of Yakutia and their importance for understanding the Quaternary history of permafrost area (abst.). Int. Union for Quat. Research (INQUA) 10th Cong., Birmingham, England, 16-24 Aug., 1977, Absts: 234.

King, Lorenz. 1979. Palsen und Permafrost in Quebec. *In* Müller-Wille, Ludger and Schroeder-Lanz, Hellmut, eds. Kanada und das Nordpolargebiet. Trierer Geog. Stud., S. *2*: 141-156.

Koaze, Takashi, Nogami, Michio and Iwata, Shuji. 1974. Paleoclimatic significance of fossil periglacial phenomena in Hokkaido, northern Japan. Quat. Research *12*(4): 177-191.

Konishchev, V.N. 1973. Kriogennoye vyvetrivaniye. *In* Akademiya Nauk SSSR, Sektsiya Nauk o Zemle, Sibirskoye Otdeleniye, II Mezhdunarodnaya Konferentsiya po Merzlotovedeniyu, Doklady i soobshcheniya *3* (Genezis, sostav i stroyeniye merzlykh tolshch i podzemnyye l'dy). Yakutsk, Yakutskoye Knizhnoye Izdatel'stvo: 38-45.

Konishchev, V.N. 1978a. Frost weathering [Kriogennoye vyvetrivaniye]. *In* Sanger, F.J., ed. Permafrost 2nd Int. Conf., Yakutsk, USSR, 13-28 July, 1973: USSR Contribution. Washington, D.C. NAS(U.S.A.): 176-181.

Konishchev, V.N. 1978b. Ustoychivost' mineralov v zone kriolitogeneza [Mineral stability in the zone of cryolithogenesis]. *In* 3rd Int. Conf. on Permafrost, Edmonton, Alta. 10-13 July, 1978. Proc. *1*. Ottawa, NRC Canada Pub. *16529*: 305-311.

Konishchev, V.N. 1980. Mineral stability in the zone of cryolithogenesis [Ustoychivost' mineralov v zone kriolitogeneza]. *In* 3rd Int. Conf. on Permafrost, Edmonton, Canada, 10-13 July, 1978. English translations of invited theme papers, Part I. Ottawa, NRC Canada, Pub. *18119*: 279-295.

Konishchev, V.N. and Kartashova, G.G. 1972. Osnovnye etapy osadkonakopleniya i razvitiya rastitel'nosti yuznoi chasti Yano-Indigirskoi nizmennosti v kainozoe. Vestnik Moskovskogo Universiteta no. *2*: 67-73.

Konrad, J.-M. and Morgenstern, N.R. 1980. A mechanistic theory of ice lens formation in fine-grained soils. Can. Geotech. J. *17*: 473-486.

Konrad, J.-M. and Morgenstern, N.R. 1981. The segregation potential of a freezing soil. Can. Geotech. J. *18*: 482-491.

Kukla, G.J. and 8 other authors. 1977. New data on climatic trends. Nature *270*: 573-580.

Lachenbruch, A.H. 1966. Contraction theory of ice wedge polygons: a qualitative discussion. *In* Permafrost Int. Conf., Lafayette, Indiana, 11-15 Nov., 1963. Proc. NAS/NRC(U.S.A.) Pub. *1287*: 63-71.

Lundqvist, Jan. 1962. Patterned ground and related frost phenomena in Sweden. Sveriges Geol. Undersökning, Årsbok *55* (1961:7) (Avh. och uppsatser, ser. C, no. *583*): 101 pp.

Lundqvist, Jan. 1969. Earth and ice mounds: a terminological discussion. *In* Péwé, T.L., ed. The periglacial environment. Montreal, McGill-Queen's Univ. Press: 203-214.

MacDonald, G.J.F. et al. 1979. The carbon dioxide problem: implications for policy in the management of energy and other resources. Exhibit 1, pp. 2-4 in Senate-Congressional symposium on CO_2 and energy policy. Congressional Record *125*(1): 1-6.

Macfadyen, W.A. 1950. Vegetation patterns in the semidesert plains of British Somaliland. Geog. J. *116*: 199-211.

Mackay, J.R. 1956. Notes on oriented lakes of the Liverpool Bay area, Northwest Territories. Rev. Can. Géographie *10*: 169-173.

Mackay, J.R. 1963. The Mackenzie Delta area, NWT. Canada Dept. Mines and Tech. Surv., Geog. Br., Mem. *8*: 202 pp.

Mackay, J.R. 1971. The origin of massive icy beds in permafrost, western arctic coast, Canada. Can. J. Earth Sciences *8*: 397-422.

Mackay, J.R. 1972. The world of underground ice. Assoc. Amer. Geographers, Ann. *62*: 1-22.

Mackay, J.R. 1973a. The growth of pingos, western arctic coast, Canada. Can. J. Earth Sciences *10*: 979-1004.

Mackay, J.R. 1973b. Problems in the origin of massive icy beds, western Arctic, Canada. *In* Permafrost 2nd Int. Conf., Yakutsk, USSR, 13-28 July, 1973: North American Contribution. Washington, D.C., NAS(U.S.A.): 223-228.

Mackay, J.R. 1974. Seismic shot holes and ground temperatures, Mackenzie Delta area, Northwest Territories. Geol. Surv. Canada, Paper *74-1A*: 389-390.

Mackay, J.R. 1975. The closing of ice-wedge cracks in permafrost, Garry Island, Northwest Territories. Can. J. Earth Sciences *12*: 1668-1674.

Mackay, J.R. 1976a. Pleistocene permafrost, Hooper Island, Northwest Territories. Geol. Surv. Canada, Paper *76-1A*: 17-18.

Mackay, J.R. 1976b. Ice segregation at depth in permafrost. Geol. Surv. Canada, Paper *76-1A*: 287-288.

Mackay, J.R. 1978. Contemporary pingos: a discussion. Biul. Peryglacjalny *27*: 133-154.

Mackay, J.R. 1979a. An equilibrium model for hummocks (nonsorted circles), Garry Island, Northwest Territories. Geol. Surv. Canada, Paper *79-1A*: 165-167.

Mackay, J.R. 1979b. Pingos of the Tuktoyaktuk Peninsula area, Northwest Territories. Géographie Phys. Quat. *33*: 3-61.

Mackay, J.R. 1980. Deformation of ice-wedge polygons, Garry Island, Northwest Territories. Geol. Surv. Canada, Paper *80-1A*: 287-291.

Mackay, J.R. 1981. Active layer slope movement in a continuous permafrost environment, Garry Island, Northwest Territories, Canada. Can. J. Earth Sciences *18*: 1666-1680.

Mackay, J.R. and Black, R.F. 1973. Origin, composition, and structure of perennially frozen ground and ground ice: a review. *In* Permafrost 2nd Int. Conf., Yakutsk, USSR, 13-28 July, 1973: North American Contribution. Washington, D.C., NAS(U.S.A.): 185-192.

Mackay, J.R., Konishchev, V.N. and Popov, A.I. 1979. Geological control of the origin, characteristics, and distribution of ground ice. *In* 3rd Int. Conf. on Permafrost, Edmonton, Canada, 10-13 July, 1978. Proc. *2*. Ottawa, NRC Canada Pub. *16529*: 1-18.

Mackay, J.R. and Lewis, C.P. 1981. Frost heave at Inuvik, N.W.T. *In* 4th Can. Permafrost Conf., Calgary, Alberta, 2-6 Mar., 1981. Absts.: 116-117.

Mackay, J.R. and Stager, J.K. 1966. The structure of some pingos in the Mackenzie Delta area, NWT. Geog. Bull. 8: 360-368.

Matthes, F.E. 1900. Glacial sculpture of the Bighorn Mountains, Wyoming. U.S. Geol. Surv., 21st Annual Rep. Pt. 2: 173-190.

Miller, R.D., Loch, J.P.G. and Bresler, E. 1975. Transport of water and heat in a frozen permeameter. Soil Sciences Soc. Amer., Proc. *39*: 1029-1036.

Miotke, F.-D. 1979. Permafrosthänge im Yukon Tanana-Upland. In Müller-Wille, Ludger and Schroeder-Lanz, Hellmut, eds. Kanada und das Nordpolargebiet. Trierer Geog. Stud. S. *2*: 112-140.

Müller, Fritz, 1959. Beobachtungen über Pingos. Detailuntersuchungen in Ostgrönland und der kanadischen Arktis. Medd. om Gronland *153*(3); 127 pp.

Müller, Fritz. 1963. Observation on pingos (Beobachtungen über Pingos). NRC Canada, Tech. Transl. *1073*: 117 pp.

National Academy of Sciences. 1974. Priorities for basic research on permafrost. Washington, D.C., NAS(U.S.A.): 54 pp.

O'Neill, Kevin and Miller, R.D. 1980. Numerical solutions for rigid-ice model of secondary frost heave. Presented at 2nd Int. Symp. on Ground Freezing, Trondheim, Norway, June, 24-26, Preprints; 656-669.

Parmuzina, O. Yu. 1978. Kriogennoye stroenie i nektorye osobenosti l'dovydeleniya v sezonnotalom sloe (The cryogenic structure and certain features of ice separation in a seasonally thawed layer). *In* Popov, A.I., ed. Problemy kriolitologii (Problems of cryolithology). Moskva, Izdatel'stvo Moskovskogo Universiteta 7: 141-163.

Parmuzina, O.Yu. 1979. K voprosu o perevaspredelenii vlagi v merzlykh gruntakh (po naturnym nablyudeniyam) [An approach to the question of the redistribution of moisture in frozen soils (according to full-scale observations)]. *In* Popov, A.I., ed. Problemy kriolitologii (Problems of cryolithology). Moskva, Izdatel'stvo Moskovskogo Universiteta 8: 194-197.

Parmuzina, O. Yu. 1980. Cryogenic texture and some characteristics of ice formation in the active layer [Kriogennoye stroenie i nektorye osobenosti l'dovydeleniya v sezonnotalom sloe]. Polar Geog. and Geol. 4(3): 131-152.

Perfect, E. and Williams, P.J. 1980. Thermally induced water migration in frozen soils. Cold Regions Science and Technology 3: 101-109.

Péwé, T.L. 1948. Origin of the Mima Mounds. Sci. Monthly 66: 293-296.

Péwé, T.L. 1966a. Ice-wedges in Alaska - classification, distribution, and climatic significance. In Permafrost Int. Conf., Lafayette, Indiana, 11-15 Nov., 1963, Proc. NAS/NRC(U.S.A.) Pub. 1287: 76-81.

Péwé, T.L. 1966b. Paleoclimatic significance of fossil ice wedges. Biul. Peryglacjalny 15: 65-73.

Péw, T.L. 1969. The periglacial environment. In Péwé, T.L., ed. The periglacial environment. Montreal, McGill-Queen's Univ. Press: 1-9.

Péwé, T.L. 1973. Ice wedge casts and past permafrost distribution in North America. Geoforum 15: 15-26.

Pissart, A. 1970. Les phénomènes physiques essentielles liés au gel, les structures périglaciaires qui en résultent et leur signification climatique. Soc. Géol. Belg. Ann. 93: 7-49.

Pissart, A. and French, H.M. 1976. Pingo investigations, north-central Banks Island, Canadian Arctic. Can. J. Earth Sciences 13: 937-946.

Plafker, George. 1964. Oriented lakes and lineaments of northeastern Bolivia. Geol. Soc. Amer., Bull. 75: 503-522.

Popov, A.I. 1978. Cryolithogenesis, the composition and structure of frozen rocks, and ground ice (The current state of the problem). Biul. Peryglacjalny 27: 155-170.

Popov, A.I. and Katasonov, E.M. 1978. Genesis, composition and structure of permafrost and ground ice. In Sanger, F.J., ed. Permafrost 2nd Int. Conf., Yakutsk, USSR, 13-28 July, 1973: USSR Contribution. Washington, D.C., NAS(U.S.A.): 713-733.

Porsild, A.E. 1938. Earth mounds in unglaciated Arctic northwestern America. Geog. Rev. 28: 46-58.

Poser, Hans. 1948. Boden- und Klimaverhältnisse in Mittel- und West-europa während der Würmeiszeit. Erdkunde 2: 53-68.

Potter, Noel, jr. 1969. Rock glaciers and mass-wastage in the Galena Creek area, northern Absaroka Mountains. Univ. Minnesota, Ph.D. Thesis: 150 pp.

Potter, Noel, jr. 1972. Ice-cored rock glacier, Galena Creek, northern Absaroka Mountains, Wyoming. Geol. Soc. Amer. Bull. 83: 3025-3057.

Railton, J.B. and Sparling, J.H. 1973. Preliminary studies on the ecology of palsa mounds in northern Ontario. Can. J. Botany 51: 1037-1044.

Rapp, Anders and Annersten, Lennart. 1969. Permafrost and tundra polygons in northern Sweden. In Péwé, T.L., ed. The periglacial environment. Montreal, McGill-Queen's Univ. Press: 65-91.

Raup, H.M. 1965. The structure and development of turf hummocks in the Mesters Vig district, Northeast Greenland. Medd. om Grønland 166(3): 113 pp.

Reger, R.D. and Péwé, T.L. 1976. Cryoplanation terraces: indicators of a permafrost environment. Quat. Research 6: 99-109.

Rein, R.G., jr. and Burrous, C.M. 1980. Laboratory measurements of subsurface displacements during thaw of low-angle slopes of a frost-susceptible soil. Arctic and Alpine Research 12: 349-358.

Revelle, Roger, et al. 1980. Environmental and societal consequences of a possible CO_2-induced climate change: a research agenda. U.S. Dept. Energy, Carbon Dioxide Research and Assessment Program 013, DOE/EV/10019-01. Uc-11. v. 1: 124 pp.

Salmi, M. 1968. Development of palsas in Finnish Lapland. 3rd Int. Peat Cong., Quebec, Canada, 19-23 Aug., 1968, Proc.: 182-189.

Salvigsen, Otto. 1977. An observation of palsa-like forms in Nordauslandet, Svalbard. Norsk Polarinst. Årsbok 1976: 364-367.

Schafer, J.P. and Hartshorn, J.H. 1965. The Quaternary of New England. In Wright, H.E., Jr. and Frey, D.G., eds. The Quaternary of the United States. Princeton, N.J., Princeton Univ. Press: 113-128.

Schunke, Ekkehard. 1975. Die Periglazialerscheinungen Islands in Abhängigkeit von Klima und Substrät. Akad. Wiss. Göttingen, Abh. III. Math.-Phys. Kl. 30: 273 pp.

Sellman, P.V. 1967. Geology of the USA CRREL permafrost tunnel, Fairbanks, Alaska. U.S. Army, Cold Regions Research and Eng. Lab., Tech. Rep. 199: 22 pp.

Sellman, P.V. 1972. Geology and properties of materials exposed in the USA CRREL permafrost tunnel. U.S. Army Corps of Engineers, Cold Regions Research and Eng. Lab., Spec. Rep. 177: 16 pp.

Semmel, Arno. 1976. Aktuelle subnivale Hang- und Talentwicklung im zentralen West-Spitsbergen. Deutscher Geographentag, Innsbruck, 1975. Verhandlungen 40: 396-400.

Sharp, R.P. 1942. Periglacial involutions in northeastern Illinois J. Geology 50: 113-133.

Shilts, W.W. 1978. Nature and genesis of mudboils, central Keewatin, Canada. Can. J. Earth Sciences 15: p. 1053-1068.

Shumskiy, P.A. 1964. Principles of structural glaciology. Translated by David Kraus. N.Y., Dover Publications: 497 pp.

Shumskiy, P.A. and Vtyurin, B.I. 1966. Underground ice. In Permafrost Int. Conf., Lafayette, Indiana, 11-15 Nov., 1963, Proc. NAS/NRC (U.S.A.) Pub. 1287: 108-113.

Solomatin, V.I. 1979. Novy dannye o stroenii i mekhanizme rosta epigeneticheskikh zhil l'da [New data on the structure and mechanism of growth of epigenetic ice wedges]. In Popov, A.I., ed. Problemy kriolitologii (Problems of cryolithology). Moskva, Izdatel'stvo Moskovskogo Universiteta: 8: 157-162.

Soons, J.M. 1962. A survey of periglacial features in New Zealand. In McCaskill, Murray, ed. Land and livelihood - Geographical essays in honor of George Jobberns. Christchurch, N.Z. Geog. Soc.: 74-87.

Stäblein, Gerhard. 1977. Periglaziale Formengesellschaften und rezente Formungsbedingungen in Grönland. In Poser, Hans, ed. Formen, Formengesellschaften und Untergrenzen in den heutigen periglazialen Höhenstufen der Hochgebirge Europas und Afrikas zwischen Arktis und Äquator. Bericht über ein Symposium. Akad. Wiss. Göttingen, Abh. III. Math.-Phys- Kl. *31*: 18-33.

Steche, Hans. 1933. Beiträge zur Frage der Strukturboden. Sächsischen Akad. Wiss. Leipzig. Berichte Math.-Phys. Kl. *85*: 193-272.

Sugden, D.E. 1977. Reconstruction of the morphology, dynamics, and thermal characteristics of the Laurentide ice sheet at its maximum. Arctic and Alpine Research *9*: 21-47.

Takagi, Shunsuke. 1980. The adsorption force theory of frost heaving. Cold Regions Science and Technology *3*: 57-81.

Thorarinsson, Sigurdur. 1951. Notes on patterned ground in Iceland, with particular reference to the Icelandic 'flás'. Geog. Annal. *33*: 144-156.

Thorn, C.E. 1974. An analysis of nivation processes and their geomorphic significance, Niwot Ridge, Colorado Front Range. Univ. Colorado, Ph.D. Thesis: 351 pp.

Thorn, C.E. 1976. Quantitative evaluation of nivation in the Colorado Front Range. Geol. Soc. Amer. Bull. *87*: 1169-1178.

Thorn, C.E. 1979. Bedrock freeze-thaw weathering regime in an alpine environment, Colorado Front Range. Earth Surface Processes *4*: 211-228.

Tumel', N.V. 1979. Temperaturnyy rezhim merzlykh porod i morozoboynoe rastreskivanie [The temperature regime of frozen rocks and the frost-caused cracking]. *In* Popov, A.I., ed. Problemy kriolitologii (Problems of cryolithology). Moskva, Izdatel'stvo Moskovskogo Universiteta *8*: 26-49.

Veillette, J.J. 1980. Nonsorted circles in cohesionless fine silty sand, north-central District of Keewatin. Geol. Surv. Canada, Paper *80-1B*: 259-267.

Wahrhaftig, Clyde and Cox, Allan. 1959. Rock glaciers in the Alaska Range. Geol. Soc. Amer. Bull. *70*: 383-436.

Walters, J.C. 1978. Polygonal patterned ground in central New Jersey. Quat. Research *10*: 42-54.

Washburn, A.L. 1956. Classification of patterned ground and review of suggested origins. Geol. Soc. Amer. Bull. *67*: 823-865.

Washburn, A.L. 1967. Instrumental observations of mass-wasting in the Mesters Vig district, Northeast Greenland. Medd. om Grønland *166*(4): 318 pp.

Washburn, A.L. 1969. Weathering, frost action and patterned ground in the Mesters Vig district, Northeast Greenland. Medd. om Grønland *176*(4): 303 pp.

Washburn, A.L. 1979. Geocryology: a survey of periglacial processes and environments. London, Edward Arnold: 406 pp.

Washburn, A.L. 1980. Permafrost features as evidence of climatic change. Earth-Science Rev. *15*(1979-1980): 327-402.

Watson, Edward. 1976. Field excursions in the Aberystwyth region, 1-10 July 1975. Biul. Peryglacjalny *26*: 79-112.

Whalley, W.B. 1974. Rock glaciers and their formation as part of a glacier debris-transport system. Univ. Reading, Dept. Geography, Geog. Papers *24*: 60 pp.

Whalley, W.B. 1979. The relationship of glacier ice and rock glacier at Grubengletscher, Kanton Wallis, Switzerland. Geog. Annal. *61A*: 49-61.

White, P.G. 1976. Some observations on the origins of rock glaciers (abst.). Geol. Soc. Amer., Absts. with Program *8*: p. 299.

White, S.E. 1976a. Is frost action really only hydration shattering? A review. Arctic and Alpine Research *8*: 1-6.

White, S.E. 1976b. Rock glaciers and block fields, review and new data. Quat. Research *6*: 77-97.

Williams, J.R. and van Everdingen, R.O. 1973. Groundwater investigations in permafrost regions of North America. *In* Permafrost 2nd Int. Conf., Yakutsk, USSR, 13-28 July, 1973: North American Contribution. Washington, D.C., NAS(U.S.A.): 435-46.

Williams, P.J. and Perfect, E. 1979. Investigation of rates of water movement through frozen soils. Carleton Univ. Geotech. Science Labs., Dept. Geography. Final Rep. for Canada Dept. Energy, Mines and Resources, Earth Physics Br., Contract 02SU-KL229-8-1562, Ser. ISU77-00269: 36 pp. + 4 appendices.

Wolfe, P.E. 1953. Periglacial frost-thaw basins in New Jersey. J. Geology *61*: 133-141.

Wolfe, P.E. 1956. Pleistocene-periglacial frost-thaw phenomena in New Jersey. N.Y. Acad. Science Trans. *18*: 507-515.

Zhestkova, T.N. 1978. Resul'taty exsperimental'nykh issledovaniy protsessa promerzaniya tonkodispersnykh gruntov. [Results of experimental studies of the freezing process in the very fine-grained soils]. *In* 3rd Int. Conf. on Permafrost Edmonton, Alta., 10-13 July, 1978. Proc. *1*. Ottawa, NRC Canada Pub. *16529*: 156-162.

Zhestkova, T.N. 1980. Results of experimental studies of the freezing process in very fine-grained soils [Resul'taty exsperimental'nykh issledovaniy protsessa promerzaniya tonkodispersnykh gruntov]. *In* 3rd Int. Conf. on Permafrost, Edmonton, Alta., 10-13 July, 1978. English translations of invited theme papers, Part I. Ottawa, NRC Canada Pub. *18119*: 119-136.

NOTES ON CONTRIBUTORS

WILLIAM H. MATHEWS is a Professor of Geological Science in the University of British Columbia and was for ten years Cominco Professor and Head of the Department. In a distinguished career he has made contributions in a wider range of subjects than any other contemporary Canadian geologist, including work in petrology, vulcanology, stratigraphy, coal geology, petroleum geology, mineral deposits, sedimentation, Quaternary geology, geology of the Tertiary era, geochronology, geomorphology, glaciology, hydrology and regional geology. He obtained his Ph.D. at the University of California at Berkeley for his study of the Garibaldi glacial-vulcanic landscape in southwestern British Columbia. Dr. Mathews is also a noted alpine explorer in British Columbia and Ross Mackay's closest research collaborator.

ALFRED JAHN is the senior periglacial geomorphologist in Europe today. He was a participant in Polish expeditions to Greenland (1937), to Spitsbergen between 1957 and 1978, and to Alaska and Iceland (1960). Although working in an expeditionary context, he encouraged the establishment of continuing experimental observations. He gained his Ph.D. from the University of Lwow in 1939 on the basis of his Greenland studies. He became Professor of Physical Geography in the University of Wroclaw in 1949 and Rector of the University in 1962. He is a member of the Polish Academy of Science and has been Chairman of the Polish Geographical Society. From 1968 to 1976 he was Chairman of the International Geographical Union Commission on Present-day Geomorphological Processes. In 1959 he was the recipient of the Médaile André Dumont of the Geological Society of Belgium. In 1969-70 he was Visiting Professor in Carleton University. His research interests include mass movement, permafrost and denudation processes in cold climates.

ANDERS RAPP has made contributions to geomorphology and to environmental science through painstaking field work in three areas: understanding of mass movement and slope stability in Sweden, Norway and Spitsbergen (1951-68); multidisciplinary and multinational soil erosion studies in Tanzania (1968-72); the 'desertification' problem in Africa (1972-76, as a member of the Secretariat for International Ecology in Stockholm). He gained his Ph.D. at the University of Uppsala in 1961 for his studies of mass wasting in Kärkevagge, an arctic-alpine valley in northern Sweden. He remained at that institution as senior lecturer until becoming Professor of Physical Geography in the University of Lund in 1976. Professor Rapp was Chairman of the International Geographical Union Commission on Field Experiments in Geomorphology between 1976 and 1980. In 1962 he was the recipient of the Kirk Bryan Award of the Geological Society of America for the work in Kärkevagge. His present research focusses on applied and basic geomorphology in northern Scandinavia.

BLAIR FITZHARRIS obtained his Ph.D. degree under the supervision of J.R. Mackay for a study of snow accumulation on the mountains near Vancouver, drawing attention to the importance of air mass trajectory on the patterns of snow accumulation and the significance of snow load for water storage. He is now a senior lecturer in Geography in the University of Otago, Dunedin, New Zealand. He has initiated studies of snow in the Southern Alps, including contribution to regional runoff, recreation potential, and avalanche hazard. Dr. Fitzharris has also researched aspects of agricultural climatology in central Otago. He returned to Canada in 1978 to work at the National Research Council on the avalanche history of Rogers Pass, British Columbia.

LORNE GOLD, a graduate in Engineering Physics, has been successively research officer, Head of the Snow and Ice Section, Head of the Geotechnical Section, Assistant Director and, since 1979, Associate Director of the Division of Building Research of the National Research Council of Canada. Since 1975 he has also been Chairman of the Associate Committee on Geotechnical Research, NRCC. He gained his Ph.D. from McGill University in 1970. In 1979 he was elected Fellow of the Royal Society of Canada. He was President of the International Glaciological Society for the period 1978-81. His research interests include deformation and failure of ice, ice engineering and ground thermal regime.

MICHAEL W. SMITH is Associate Professor of Geography in Carleton University, Ottawa. His research as a graduate student on permafrost aggradation was supervised by J. R. Mackay, with field study in the Mackenzie Delta. The behaviour of freezing ground has provided the focus of his subsequent research: he has given attention to arctic pipelines, to effects on permafrost of surface disturbance, and to methods for measuring moisture content in freezing ground. The Canada Department of Environment and Canada Department of Energy, Mines and Resources have supported phases of the work. With other members of the Carleton University Geotechnical Science Laboratory staff, Dr. Smith participates in an exchange with the Ecole des Ponts et Chaussées in Paris.

SAMUEL I. OUTCALT is a Professor in the Department of Geology, University of Michigan, Ann Arbor, U.S.A. His doctoral research on the conditions for the inception and growth of needle ice was conducted under J. R. Mackay. He has subsequently applied his expertise in the computation of surface climate to consideration of ground thermal regime, lake circulation and the behaviour of soil fauna. However, his central interest remains the study of frozen ground, with field activity principally in northern Alaska. He maintains a working association with the United States Army Cold Regions Research and Engineering Laboratory in Hanover, New Hampshire.

NORBERT R. MORGENSTERN obtained his Ph.D. from the University of London, where he was also Lecturer in Civil Engineering in the Imperial College of Science and Technology between 1960 and 1968. In 1968 he was appointed Professor of Civil Engineering in the University of Alberta. Professor Morgenstern's research interests include the mechanics of landslides and other geomorphological processes on slopes, the properties of natural materials, permafrost engineering, underground excavation, and the design and construction of dams. He was a prize winner of the British Geotechnical Society in 1961 and 1966, Distinguished Lecturer for the Associate Committee on Geotechnical Engineering, National Research Council of Canada, in 1967, recipient of the Huber Research Prize of the American Society of Civil Engineers in 1972 and of the Canadian Geotechnical Society Research Prize in 1977, and winner of the Robert F. Legget Award of the Geological Society of America in 1979. Professor Morgenstern delivered the Rankine Lectures to the British Geotechnical Society in 1981. He is a Fellow of the Royal Society of Canada.

NIKOLAI N. ROMANOVSKIJ is Professor in the Faculty of Geology, Moscow State University, Moscow, U.S.S.R. Professor Romanovskij is an authority on ice-wedge polygons and other features induced by frost cracking in permafrost and in non-permafrost environments. He has carried out field studies in Siberia for more than twenty years, where he has also investigated groundwater icings, especially from deep sub-permafrost drainage, rock streams, and the genesis of permafrost as related to tectonic movements. He has worked in Poland on a fellowship in the Geographical Institute of the University of Lodz, and has travelled abroad, including a visit to the Canadian arctic.

A. LINCOLN WASHBURN is one of the founders in North America of the systematic study of periglacial regions and of geocryology. He was a member of the National Geographical Society's 1936 Mount McKinley Expedition, and of the Boyd East Greenland Expedition in 1937. He undertook field studies in the central Canadian Arctic during 1938-41, and again in 1949, then conducted observations at Mesters Vig, Northeast Greenland from 1955 until 1964. He obtained his Ph.D. at Yale University in 1942 and became the first executive director of the Arctic Institute of North America (1945-51). He was director of the United States Army Corps of Engineers Snow, Ice and Permafrost Research Establishment (the forerunner of CRREL) in 1952-53, then Professor of Northern Geology in Dartmouth College until 1960. Dr. Washburn then returned to Yale as Professor of Geology, becoming Faculty Professor in 1976. He was also the founding director of the Quaternary Research Center in the University of Washington (1967-75) and founding editor of *Quaternary Research*. He received the Kirk Bryan Award of the Geological Society of America in 1971 (for the work at Mesters Vig) and the Médaille André Dumont of the Geological Society of Belgium in 1973. In 1981 he returned to Resolute Bay in the Canadian Arctic to establish a new field programme.

AUTHOR CITATION INDEX.

This index lists the first page in each chapter upon which a paper or communication by the listed author is cited. Self-citations by the chapter writer are not, however, included, since the reader may directly examine the reference list in each chapter to find papers by the writer.

INDEX

Abisko (Sweden): climate, 42; studies of mass movement by A. Rapp, 38ff
Active layer: depth at Arfersiorfik Fiord, Greenland, 22; depth at Hornsund Fiord, Spitsbergen, 24; depth, influence of slope inclination on, 26; effect of surface cover, 25; in Siberia, 25; oxidation, 25; seasonal development, 22
Air intrusion value (of soil water capillary tension): 103
Ajaura (Sweden): debris slides, 41
Alaska: permafrost age, 187
Andöya (Norway): debris slide, 42
Arfersiorfik Fiord (Greenland): periglacial studies by A. Jahn, 21-22
Arjeplog (Sweden): debris slide, 42, 45
Avalanche (snow): coefficient of sliding friction, 63, 69-71; coefficient of turbulent friction, 63, 69-70; definition, 39; "dirty," 39; mass to drag ratio, 64, 70-71; maximum runout, 59; mixed, 39; occurrence in New Zealand, 68; runout distance, 57-72
Avalanche, slush: 39, 48-53; definition, 39; frequency in northern Scandinavia, 50; occurrence in northern Scandinavia, 53
Avalanche boulder tongue: 50
Avalanche debris tail: 50
Avalanche hazard: 58-59; zoning, 59
Avalanche impact features: 50
Avalanche model: Colorado regression, 63, 68-69; Norwegian, 62, 68; numerical, 64-65, 70-71; Voellmy, 63, 69-70
Avalanche path: in New Zealand, 61, 66; vegetation in, 62
Avalanche track: *see* Avalanche path

Bac Motometer: 28-29
"Bedstead" device: for measuring frost heave, 29
Block field: 164, 185; in southern Yakutia, 164
Block slope: 164, 185; in southern Yakutia, 164
Block stream: 164, 185; in southern Yakutia, 164

Capillary tension: air intrusion value, 103; effect on ice segregation, 91
Capillary (water) pressure in soil: *see* Capillary tension

Carbon dioxide: climatic effect at high latitude, 189-90
Castor silt loam: thermal conductivity at freezing temperatures, 101
Circles (periglacial landform): 172
Clapeyron equation: 101, 105ff, 114
Climatic optimum: Holocene, effect on formation of periglacial features in Siberia, 162
Cryoplanation terrace: 184; occurrence, 186-87
Cryosuction: 102

Debris flow: definition, 39; occurrence in northern Scandinavia, 40-46
Debris flow cone: 47
Debris flow levee: 39-40, 42, 45
Debris slide: definition, 39; occurrence in northern Scandinavia, 40-45
Denudation rate: by debris slides and flows in northern Scandinavia, 38, 49; by slopewash in Spitsbergen, 32
Disjoining pressure: in water film, 85

Effective stress: *see* soil: effective stress
Environmental reconstruction: in periglacial regions, 190
Erosion: slopes in Scandinavia, 38; slopes in Spitsbergen by sheetwash, 32-33

Fairbanks silt: frozen creep behaviour, 136
Fort Norman (N.W.T., Canada): periglacial slope stability investigations by N. Morgenstern, 148
Freezing: pressure developed during, 78
Freezing front: 79
Freezing soil: *see* Soil: freezing
Frequency: of alpine debris flows in northern Scandinavia, 39ff; of mass wasting in northern Scandinavia, 38; of rainfall in Spitsbergen, 33; of slush avalanches in northern Scandinavia, 50
Frost: ice segregation, 124; normal, 124; normal, Arakawa model, 125
Frost action: *see* Ice segregation: frost heave
Frost circles: 172
Frost cracking: 171-72
Frost creep: 183